THE SCIENCE OF WATER

The
SCIENCE
OF WATER

Concepts & Applications

Frank R. Spellman, Ph.D.

TECHNOMIC
PUBLISHING CO., INC.
LANCASTER · BASEL

The Science of Water
a TECHNOMIC®publication

Published in the Western Hemisphere by
Technomic Publishing Company, Inc.
851 New Holland Avenue, Box 3535
Lancaster, Pennsylvania 17604 U.S.A.

Distributed in the Rest of the World by
Technomic Publishing AG
Missionsstrasse 44
CH-4055 Basel, Switzerland

Printed in the United States of America
10 9 8 7 6 5 4 3 2 1

Main entry under title:
   The Science of Water: Concepts and Applications

A Technomic Publishing Company book
Bibliography: p.
Includes index p. 233

Library of Congress Catalog Card No. 97-62221
ISBN No. 1-56676-612-5

To Linda R. Lindenmuth, Woodhaven Water Company,
Quinton, Virginia. Linda is an attractive lady with
whom I share a bond: We appreciate and respect
the environment and the liquid substance that is
a major part of it.

Table of Contents

Preface

WATER, Water, Water—Water everywhere, right? Moreover, since water is everywhere, it is also plentiful and there is no way we will ever lack for it, right? In addition, the earth's supply of finite water resources can be increased constantly to meet growing demand, right? In spite of these absurdities, a belief prevails that the earth's finite water resources can be increased constantly to meet growing demands. At the present time, the supply of water is constantly made to respond to demand. History has demonstrated that consumption and waste increase in response to rising supply. The fact of the matter is freshwaters are a *finite* resource that can be increased only slightly through desalinization or some other practice, all at tremendous cost.

This text deals with the essence of water; that is, what water is and what water is all about. Further, while this text points out that water is one of the simplest and most common chemical compounds on earth, it also shows water to be one of the most mysterious and awe-inspiring substances we know. Important to this discussion about water and its critical importance on earth is man—man and his use, misuse, and reuse of fresh water and wastewater. Furthermore, this text takes the view that since water is the essence of all life on earth, it is precious—too precious to abuse, misuse, and ignore. Thus, as you might guess, the common thread that is woven throughout the fabric of this presentation is water resource utilization and its protection.

Written primarily as an information source, it should be pointed out that this text is not limited in its potential for other uses. For example, while this work can be utilized by the water/wastewater practitioner to provide insight into the substance he/she works hard to collect, treat, and supply for its intended purpose, it can just as easily provide important information for the policymaker who may be tasked with making decisions concerning water resource utilization. Consequently, this book will serve a varied audience: students, lay personnel, regulators, technical experts, attorneys, business leaders, and concerned citizens.

The question becomes: Why a text on the science of water? Which leads to another question: Isn't it true that water treatment, wastewater treatment, and other work with water is more of an art than a science? In answering the first question it should be pointed out that the study of water is a science. It is a science that is closely related/interrelated to other scientific disciplines such as biology, microbiology, chemistry, mathematics, hydrology, and others. Therefore, to solve the problems and understand the issues related to water, water practitioners need a broad base of scientific information from which to draw.

In answering the second question it might be easier to bring up another question or situation. Consider, for example, the thoracic surgeon (thoracic surgery is *the* major league of surgery, according to a thoracic surgeon I know), who has a reputation as being an *artist* with a scalpel. This information might be encouraging to the would-be patient who is to be operated on by such a surgeon. However, this same patient might further inquire about the surgeon's education, training, experience; about her knowledge of the science of medicine. If I were the patient, I would want her (the surgeon) to understand the science of my heart and other vital organs before she took scalpel in hand to perform her artful surgery. Wouldn't you?

The point is that if water and wastewater operators are expected to treat their media correctly, to the point where we can safely drink the treated water and the treated effluent will not damage its receiving body, then shouldn't we expect them to understand the processes they are tasked to "operate"?

Thus, for those who need to study water because a knowledge of water is integral to their professions, and/or for those who actually work with or around water, and/or for those who have an insatiable curiosity and are driven to learn something about water, this text is for you—all of you. Further, what this text is designed to do is to fill that gap in water information that has been out there (in academia as well as the real world) for some time. This lofty goal is accomplished via a user-friendly format where all unnecessary technical jargon is avoided. *The Science of Water* is presented in a straightforward, informative, thought-provoking manner; this text, in an unfettered way, puts water on that lofty pedestal where it belongs.

Moreover, the reader will quickly determine that the author presents water in a light that is nontraditional; that is, water is, in the author's view, truly the nectar of life. Moreover, this text views wastewater differently than most: Simply stated, if water is the nectar of life, then wastewater is, because of its potential for beneficial reuse, the elixir that can provide a sensible alternative toward a sustainable future.

The bottom line: This text is an adventure; it serves up the big picture while serving as a launching point for further knowledge, insight and exploration.

When you get right down to it, isn't this what reading and learning should be all about?

Acknowledgements

I would like to extend my appreciation to those who are really responsible for this publication; they provided me with more support than I can weigh: Joseph Eckenrode, Susan Farmer, Kimberly Martin, Teresa Wiegand, and Nancy Whiting—all from Technomic Publishing Company. Thank you—very much!

CHAPTER 1

Introduction

Whether we characterize it as soft summer rain, as fog, as flood or avalanche, or as stimulating as a stream or cascade, water is special—water is strange—water is different.

Water is the most abundant inorganic liquid in the world; moreover, it occurs naturally anywhere on earth. Literally awash with it, life depends on it, and yet water is so very different.

*Water is scientifically different. With its rare and distinctive property of being denser as a liquid than as a solid, it is different. Water is different in that it is the only chemical compound found naturally in solid, liquid, or gaseous states. Water is sometimes called the **universal solvent**. This is a fitting name, especially when you consider that water is a powerful reagent, which is capable in time of dissolving everything on earth.*

Water is different. It is usually associated with all the good things on earth. For example, water is associated with quenching thirst, with putting out fires, and with irrigating the earth. The question is: Can we really say emphatically, definitively that water is associated with only those things that are good?

Not really!

Remember, water is different; nothing, absolutely nothing, is safe from it.

Water is different. This unique substance is odorless, colorless, and tasteless. Water covers 71% of the earth completely. Even the driest dust ball contains 10–15% water.

Water and life—life and water—inseparable.

Three hundred twenty-six million cubic miles of water cover earth but only 3% of this total is fresh with most locked up in polar ice caps, in glaciers, in lakes, in flows through soil and in river and stream sys-

1

tems back to an ever increasingly saltier sea (only 0.027% is available for human consumption). Water is different.

Salt water is different from fresh water. Moreover, this text deals with fresh water and ignores salt water because salt water fails its most vital duty, which is to be pure, sweet, and serve to nourish us.

Water is special—water is strange—water is different—more importantly, water is critical to our survival, yet we abuse it, discard it, pollute it, curse it, dam it, and ignore it. At least this is the way we view the importance of water at this moment in time . . . however, because water is special, strange, and different, the dawn of tomorrow is pushing for quite a different view.

1.1 SETTING THE STAGE

ALONG with being special, strange, and different, water is fascinating. It is fascinating for several reasons. For instance, one only need review a few facts about water to gain appreciation for its uniqueness: Water exists as a liquid between 0°C and 100°C (32°F and 212°F); exists as a solid at or below 0°C (32°F); and exists as a gas at or above 100°C (212°F). One gallon of water weighs 8.33 pounds (3.778 kilograms). One gallon of water equals 3.785 liters. One cubic foot of water equals 7.50 gallons (28.35 liters). One ton of water equals 240 gallons. One acre foot water equals 43,560 cubic feet (325,900 gallons). Earth's rate of rainfall equals 340 cubic miles per day (16 million tons per second). Finally, the fact is water is dynamic (constantly in motion), evaporating from sea, lakes, and the soil; being transported through the atmosphere; falling to earth; running across the land; and filtering downward to flow along rock strata.

Consider another fact about water: The availability of a water supply adequate in terms of both quantity and quality is *essential* to our very existence. One thing is certain: History has shown that the provision of an adequate quantity of quality potable water has been a matter of major concern since the beginning of civilization.

Water—especially clean water—we know we need it to survive—we know a lot about it—however, the more we know the more we discover we don't know.

Modern technology has allowed us to tap potable water supplies and to design and construct elaborate water distribution systems. Moreover, we have developed technology to treat water we foul, soil, pollute, discard, and flush away.

Have you ever wondered where the water goes when you flush the toilet? Probably not.

An entire technology has developed around treating water and wastewater. Along with technology, of course, technological experts have been developed. These experts range from environmental/structural/civil engineers to environmental scientists, geologists, hydrologists, chemists, biologists, and others.

Along with those who design and construct water/wastewater treatment works, there is a large cadre of specialized technicians, spread worldwide, who operate water and wastewater treatment plants. These operators are tasked, obviously, with either providing a water product that is both safe and palatable for consumption and/or with treating (cleaning) a waste stream before it is returned to its receiving body (usually a river or stream).

With all the time and effort expended in training those who design and operate our water and wastewater systems you would just naturally assume that these folks have a very strong background and understanding of the science of water.

The fact that most of these folks do know more about water than the rest of us comes as no surprise. For the average person, knowledge of water usually extends to knowing no more than that water is good or bad; it is terrible tasting, just great, wonderful, clear and cool and sparkling, or full of scum/dirt/rust, great for the skin or hair, very medicinal, and so on. Thus, to say the water "experts" know much more about water than the average person is an accurate statement.

At this point you are probably asking yourself: What does all this have to do with anything? Good question.

What it has to do with water is quite simple. We do not know as much as we need to know about water.

As a case in point, consider this: Have you ever tried to find a text that deals exclusively and extensively with the science of water? Such texts are few, far-flung, imaginary, non-existent—there is a huge gap out there.

Then the question shifts to—why would you want to know anything about water in the first place? Another good question.

This text makes an effort to answer this question.

To start with, let's talk a little about water.

When the average person goes to the potable water tap in the sink, or to the refrigerator spigot that dispenses cool, refreshing water, or to that bottle containing that special, exotic blend of ingredients that makes up "designer or miracle water," he/she probably gives little thought to what he or she is doing; that is, drinking a glass of water.

Let's face it, drinking a glass of water is something that normally takes little effort and even less thought.

The situation could be different, however. For example, consider the young woman who is an adventurer; an outdoorsperson. She likes to jump into her four-wheel-drive vehicle and head for new adventure. On this particular day she decides to drive through Death Valley, California—from one end to the other and back. During her transit of this isolated desert region, she decides to take a side road that seems to lead to the mountains to her right.

She travels along this isolated, hardpan road for approximately 50 miles—then the motor in her four-wheel-drive vehicle suddenly quits. No matter what she does, the vehicle will not start. Eventually, the vehicle's battery dies; she had cranked on it too much.

Realizing that the vehicle was not going to start, she also realized she was alone and deep inside an inhospitable area. What she did not know was that the nearest human being was about 60 miles to the west.

She had another problem—a problem more pressing than any other. She did not have a canteen or container of water. Obviously, she told herself, this is not a good situation.

What an understatement this turned out to be.

Just before noon she started back down the same road she had traveled. She reasoned she did not know what was in any other direction other than the one she had just traversed. She also knew the end of this side road intersected the major highway that bisected Death Valley. She could flag down a car or truck; she would get help, she reasoned.

She walked—and walked—and walked some more. "Gee, if it wasn't so darn hot," she muttered to herself, to sagebrush, to scorpions, to rattlesnakes and to cacti. The point is it was hot; about 107°F.

She continued on for hours, but now she was not really walking; instead, she was forcing her body to move along. Each step hurt. She was burning up. She was thirsty. How thirsty was she? Well, right about now just about anything liquid would do, thank you very much!

Later that night, after hours of walking through that hostile land, she couldn't go on. Deep down, in her heat-stressed mind, she knew she was in serious trouble. Trouble of the life-threatening variety.

Just before passing out, she used her last ounce of energy to issue a dry pathetic scream.

This scream of lost hope and imminent death was heard—but only by the sagebrush, the scorpions, the rattlesnakes, and the cacti—and by the vultures that were now circling above her parched, dead remains. The vultures were of no help, of course. They had heard these screams before. They were indifferent; they had all the water they needed; their food supply wasn't all that bad either.

The preceding case sheds light on a completely different view of water. Actually, it is a very basic view that holds: We cannot live without it.

If water is so precious, so necessary for sustaining life, then two questions arise: (1) Why do we ignore water? (2) Why do we abuse it (pollute or waste it)?

We ignore water because it is so common, so accessible, so available, so unexceptional (unless you are lost in the desert without a supply of it). Why do we pollute and waste water? There are several reasons; many will be discussed later in this text.

You might be asking yourself: Is water pollution really that big of a deal? Simply stated, yes; it is.

As this text proceeds it will lead you down a path strewn with abuse and disregard for our water supply—then all (excepting the water) will become clear. In the meantime, it might be helpful to review the following example of man's abuse and disregard for water (in this case the water of a major ocean).

If you have ever transited from the East Coast of the United States to Europe via ship across the Atlantic Ocean, you probably travelled the "Great Circle Route." This Great Circle Route is the shortest distance between the two continents. This is the case because the route follows the circular geometry of Earth.

During this transit, the ship's captain uses Satellite Navigation Aids (SatNav), other electronic aids, and standard ship's compass. Obviously, this is practicing good seamanship (and a lot of common sense).

During a transit of this same route, a few years ago, when I had the distinct honor and pleasure of performing officer-of-the-deck duties on a U.S. Navy warship, one thing became quite apparent to me almost instantly. I found out that if the ship's SatNav, electronic navigation aids, and compass failed, and if the sky was overcast for the entire voyage (this would not allow me to navigate by using the stars or sunlines, etc.), I would have little difficulty in finding my way, in navigating the ship to its European destination.

I would have had little difficulty in navigating because all I would have had to do was to follow the trail—the markers—the signposts—the remnants of previous ships following the same path. Never had I seen such a clearly marked seaway. Garbage, refuse of all descriptions, and plastic-like debris floating along. This floating trail was narrow but distinct. At night this garbage trail could easily be seen in the light shone from the ship's dim lights.

When I first saw this garbage trail, I just shook my head in disgust. There was no doubt whatsoever this route had been heavily travelled by others. This was apparent because man had left his footprint—man has a bad habit of doing this. What it really comes down to is "out of sight out of mind" thinking. Or when we abuse our resources in such a manner, maybe we think to ourselves: "Why worry about it. Let God sort it out."

This is exactly what worries a lot of people (including me); in the end God may do a lot of sorting out.

Getting back to that gap in knowledge dealing with the science of water. This text is designed to show how this obvious and unsatisfactory gap in information about water is to be filled in. Having said this, now its time to welcome you to the gap-filler: *The Science of Water: Concepts and Applications*.

Finally, before moving on with the rest of the text, it should be pointed out the view held throughout this work is that water is special, strange, and different—and very vital. This view is held for several reasons, but the most salient factor driving this view is the one that points to the fact that on this planet, *water is life.*

1.2 SCOPE OF TEXT

Preamble: Science is any systematic field of study or body of knowledge that aims, through experiment, observation, and deduction, to produce reliable explanation of phenomena, with reference to the material and physical world.

It is the intent of this text to abide by the precepts presented in this preamble. Moreover, this text presents the science of water in a logical, step-by-step, plain English, user-friendly format.

The text consists of twelve chapters. The introduction is presented in Chapter 1. Chapter 2 is all about water. Chapter 3 presents a thorough discussion of water mathematics (the type of math that is used when working in water/wastewater treatment). Chapter 4 discusses basic water hydraulics. Chapter 5 deals with water chemistry. Water chemistry is important because we usually think water is simple; this is the case because we rarely have to think about it in detail, or at all. The point is in order to gain even a basic understanding of water one must understand the water molecule and the chemistry of water. Chapter 6 presents water biology. Chapter 7 discusses water ecology. Water ecology is an important branch of water science that is often overlooked—but not in this text. Chapter 8 involves information related to water quality. Remember, water can be readily available but if its quality is unsuitable for human consumption, it is of little value to us. In Chapter 9, water and wastewater treatment is discussed. Chapter 10 presents information on water use. Chapter 11 deals with water pollution. Chapter 12, the final part, discusses water reuse. The subject matter contained in these twelve parts is mixed, blended, and collated in such a manner so as to provide an easy read that serves as a ready reference and source of information on the science of water.

Having said all this, it should be pointed out that this text is specifically written to serve as a reference text for an undergraduate course and also as a guide to prepare water/wastewater specialists for licensing examinations. It also serves as a basic primer for those who are interested in gaining knowledge about water-related topics, for whatever reason. The text follows a pattern that is nontraditional; that is, the paradigm (model or proto-type) used here is based on real-world experience—not on theoretical gob-bledygook. It is the knowledge gained from work performed in the world of water and wastewater treatment that this text is all about.

1.3 DEFINITIONS OF KEY TERMS

Every branch of science, including water science, has its own language for communication. The terminology used herein is as different from that of astronomy as that of sanitary engineering is from aeronautical engineering. In order to work even at the edge of water science and the science disciplines closely related to water science, it is necessary for the reader to acquire a familiarity with the vocabulary used in this text.

While it is helpful and important for technical publications to include definitions or a glossary of key terms at the end of the work, for the reader's use, it is more important, in my view, to include many of these key definitions early in the text to facilitate a more orderly, logical, step-by-step learning activity. Thus, in the following section some of the key terms are listed and defined. Other terms not defined here will be defined when they are used in the text.

1.3.1 DEFINITIONS

- *Absolute pressure* the total pressure in a system, including both the pressure of water and the pressure of the atmosphere (about 14.7 psi, at sea level).
- *Absorption* assimilation of molecules or other substances into the physical structure of a liquid or solid without chemical reaction.
- *Acid* any substance that releases hydrogen ions (H^+) when it is mixed into water.
- *Acidic solution* a solution that contains significant numbers of (H^+) ions.
- *Adsorption* physical adhesion of molecules or colloids to the surfaces of solids without chemical reaction.
- *Aerobic organism* an organism that requires oxygen for respiration.
- *Algae* simple plants containing chlorophyll. Many are microscopic, but under conditions favorable for their growth they grow in colonies and produce mats and similar nuisance masses (algal blooms).
- *Alkaline solution* a solution that contains significant numbers of (OH^-) ions. A basic solution.

- *Alkalinity*   a measurement of water's capacity to neutralize an acid.
- *Anaerobic organism*   one that can thrive in the absence of oxygen.
- *Anion*   a negatively charged ion resulting from dissociation of salts, acids, or alkalies in aqueous solution.
- *Aquifer*   a porous, water-bearing geologic formation.
- *Arithmetic mean*   a measurement of average value, calculated by summing all terms and dividing by the number of terms.
- *Arithmetic scale*   a scale is a series of intervals (marks or lines), usually marked along the side and bottom of a graph, that represents the range of values of the data. When the marks or lines are equally spaced, it is called an arithmetic scale.
- *Atom*   the smallest particle of an element that still retains the characteristics of that element.
- *Atomic number*   the number of protons in the nucleus of an atom.
- *Atomic weight*   the sum of the number of protons and the number of neutrons in the nucleus of an atom.
- *Average flow rate*   the average of the instantaneous flow rates over a given period of time, such as a day.
- *Bacteria*   microscopic, single-cell plants that reproduce by fission or by spores, identified by their shapes: coccus, spherical; bacillus, rod-shaped; and spirillum, curved.
- *Base*   any substance that releases hydroxyl ions (OH^-) when it dissociates in water.
- *Biome*   the largest land community unit which is convenient to recognize (Odum, 1971).
- *Biota*   all living organisms of a region or system.
- *BOD*   biochemical oxygen demand of a water, being the oxygen required by bacteria for oxidation of the soluble organic matter under controlled test conditions.
- *Buffer*   a substance capable in solution to resist a reduction in pH as acid is added.
- *Carbonate hardness*   caused primarily by bicarbonate.
- *Cation*   a positively charged ion resulting from dissociation of molecules in solution.
- *Chemical bond*   the force that holds atoms together within molecules. A chemical bond is formed when a chemical reaction takes place. Two types of chemical bond are ionic bonds and covalent bonds.
- *Chemical reaction*   a process that occurs when atoms of certain elements are brought together and combine to form molecules, or when molecules are broken down into individual atoms.
- *Coagulation*   the neutralization of the charges of colloidal matter.

- *COD*   chemical oxygen demand, a measure of organic matter and other reducing substances in water.
- *Coliform bacteria*   those found in the intestinal tract of warm-blooded animals; used as indicators of pollution if found in water.
- *Concentration*   in chemistry, a measurement of how much solute is contained in a given amount of solution. Concentrations are commonly measured in milligrams per liter (mg/L).
- *Covalent bond*   a type of chemical bond in which electrons are shared.
- *Density*   the weight of a substance per unit of its volume; e.g., pounds per cubic foot or pounds per gallon.
- *Detention time*   the average length of time a drop of water or a suspended particle remains in a tank or chamber. Mathematically, it is the volume of water in the tank divided by the flow rate through the tank.
- *Disinfection*   application of energy or chemicals to kill pathogenic organisms.
- *Dynamic discharge head*   the difference in height measured from the pump center line at the discharge of the pump to the point on the hydraulic grade line directly above it.
- *Dynamic suction head*   the distance from the pump center line at the suction of the pump to the point of the hydraulic grade line directly above it. Dynamic suction head exists only when the pump is below the piezometric surface of the water at the pump station. When the pump is above the piezometric surface, the equivalent measurement is dynamic suction lift.
- *Dynamic suction lift*   the distance from the pump center line at the suction of the pump to the point on the hydraulic grade line directly below it. Dynamic suction lift exists only when the pump is above the piezometric surface of the water at the pump suction. When the pump is below the piezometric surface, the equivalent measurement is called dynamic suction head.
- *Effective height*   the total feet of head against which a pump must work.
- *Element*   any of more than 100 fundamental substances that consist of atoms of only one kind and that constitute all matter.
- *Elevation head*   also called position or potential head; the energy possessed per unit weight of a fluid because of its elevation above some reference point.
- *Eutrophication*   enrichment of water, causing excessive growth of aquatic plants and an eventual choking and deoxygenation of the water body.

- *Facultative organisms*   microbes capable of adapting to either aerobic or anaerobic environments.
- *Flow rate*   a measure of the volume of water moving past a given point in a given period of time.
- *Flume*   a raceway or channel constructed to carry water or to permit measuring its flow.
- *Friction head loss*   the head lost by water flowing in a stream or conduit as the result of (1) the disturbance set up by the contact between the moving water and its containing vessel and (2) intermolecular friction.
- *Fungi*   as applied to water, these are simple, one-celled organisms without chlorophyll, often filamentous.
- *Hardness*   the concentration of calcium and magnesium salts in water.
- *Head*   a measure of the energy possessed by water at a given location in the water system, expressed in feet. Also, it is a measure of the pressure or force exerted by water, expressed in feet.
- *Head loss*   the amount of energy used by water in moving from one location to another.
- *Hydraulic gradient*   established by the slope of the water table which causes groundwater to flow.
- *Ion*   an atom or radical in solution carrying an integral electrical charge either positive (cation) or negative (anion).
- *Ionic bond*   a type of chemical bond in which electrons are transferred.
- *Langelier saturation index*   a means of expressing the degree of saturation of a water as related to calcium carbonate solubility.
- *Mixture*   two or more elements, compounds, or both, mixed together with no chemical reaction occurring.
- *Molarity*   a measure of concentration defined as the number of moles of solute per liter of solution.
- *pH*   a means of expressing hydrogen ion concentration in terms of the powers of 10; a measurement of how acidic or basic a substance is. The pH scale runs from 0 (most acidic) to 14 (most basic). The center of the range (7) indicates the substance is neutral.
- *Piezometric surface*   an imaginary surface that coincides with the level of water in an aquifer, or the level to which water in a system would rise in a piezometer (AWWA, 1995a).
- *Plankton*   small organisms with limited powers of locomotion, carried by water currents from place to place.
- *Pollutant*   a contaminant at a concentration high enough to endanger the aquatic environment.

- *Pressure* the force pushing on a unit area. Normally, in water applications, measured in feet of head.
- *Pressure head* a measurement of the amount of energy in water due to water pressure.
- *Protozoa* large, microscopic single-cell organisms higher on the food chain than bacteria, which they consume.
- *Solute* the substance dissolved in a solution.
- *Solution* a liquid containing a dissolved substance.
- *Specific gravity* the ratio of the density of a substance to a standard density. For solids and liquids, the density is compared with the density of water. For gases, the density is compared with the density of air.
- *Static discharge head* the difference in height between the pump center line and the level of the discharge free water surface (i.e., the surface of the water that is in contact with the atmosphere).
- *Static suction head* the difference in elevation between the pump center line and the free water surface of the reservoir feeding the pump.
- *Static suction lift* the difference in elevation between the pump center line of the pump and the free water surface of the liquid being pumped.
- *Thermocline* the layer in a lake dividing the upper, current-mixed zone, from the cool, lower stagnant zone.
- *Titration* a chemical test method using a reagent that is chemically keyed to the unknown factor being tested. Usually an indicator is added to the test sample and then the titrating solution is added in measured amounts. When sufficient titrating solution is added to bring about a color change in the sample (the endpoint), the amount of solution added is noted and its volume is equivalent to the factor being tested. Titration methods employ accurately calibrated burettes or Direct Reading Titrators or may follow drop count procedures.
- *Turbidity* a suspension of fine particles that obscures light rays but requires many days for sedimentation because of small particle size.
- *Turnover* the mixing of lower and upper layers in a lake in spring and fall caused by temperature and density equalization.
- *Water table* the top of the zone of saturation in the ground.

CHAPTER 2

All about Water

Water can both float and sink a ship.

2.1 INTRODUCTION

WHEN you fill a glass with water and then drink the water, has it ever occurred to you that you are drinking the same water that Leonardo da Vinci, Cleopatra, Julius Caesar, Michelangelo, Napoleon, and several billion other folks have drunk over the years? It's true.

It's true because the total quantity of water on earth is much the same now as it was more than three or four billion years ago, when the 320+ million cubic miles of it were first formed. Ever since then, the water *reservoir* has gone round and round, building up, breaking down, cooling, and then warming. Water is very durable, but remains difficult to explain, because it has never been isolated in a completely undefiled state.

Remember, water is special, strange, and different.

2.2 HOW SPECIAL, STRANGE, AND DIFFERENT IS WATER?

Have you ever wondered what the nutritive value of water is? Well, the fact is water has no nutritive value. It has none; yet it is the major ingredient of all living things.

Water is used to fight forest fires. Yet we use water spray on coal in a furnace to make it burn better.

Chemically, water is hydrogen oxide. It turns out, however, on more advanced analysis to be a mixture of more than thirty possible compounds. In addition, all of its physical constants are abnormal (strange).

At a temperature of 2,900°C some substances that contain water cannot be forced to part with it. And yet others that do not contain water will liberate it when even slightly heated.

When liquid, water is virtually incompressible; as it freezes, it expands by an eleventh of its volume.

For the above stated reasons, and for many others, we can truly say that water is special, strange, and different.

2.3 THE PROPERTIES AND CHARACTERISTICS OF WATER

Thus far we have said many things about water; however, we have not said that water is plain. This is the case because nowhere in nature is plain water to be found. Here on earth, with a geologic origin dating back over three to five billion years, water found in even its purest form is composed of many constituents. Along with H_2O molecules, hydrogen (H^+), hydroxyl (OH^-), sodium, potassium, and magnesium, there are other ions and elements present. Additionally, water contains dissolved compounds including various carbonates, sulfates, silicates, and chlorides. Rain water, often assumed to be the equivalent of distilled water, is not immune to contamination that is collected as it descends through the atmosphere. The movement of water across the face of land contributes to its contamination, taking up dissolved gases, such as carbon dioxide and oxygen, and a multitude of organic substances and minerals leached from the soil. Don't let that crystal clear lake or pond fool you. These are not filled with water alone but are composed of a complex medium of chemical ingredients far exceeding the brief list presented here; it is a special medium in which highly specialized life can occur.

All forms of life need water to survive. As a matter of fact, all forms of life have some water content. For example, the human body is more than 70% water; other animals and plants range from 50–97% water.

How important is water to life? To answer this question all we have to do is to take a look at the common cell; it easily demonstrates the importance of water to life.

Living cells comprise a number of chemicals and organelles within a liquid substance, the cytoplasm, and the cell's survival may be threatened by changes in the proportion of water in the cytoplasm. This change in proportion of water in the cytoplasm can occur through desiccation (evaporation), oversupply, or the loss of either nutrients or water to the external environment. A cell that is unable to control and maintain homeostasis (i.e., the correct equilibrium/proportion of water) in its cytoplasm may be doomed—it may not survive.

2.3.1 INFLAMMABLE AIR + VITAL AIR = WATER

In 1783, in England, Henry Cavendish (a brilliant chemist and physicist) was "playing with" electric current. Specifically, Cavendish was passing electric current through a variety of substances to see what happened. Eventually, he got around to water. He filled a tube with water and sent his current through it. The water vanished.

To say that Cavendish was flabbergasted by the results of this experiment would be to make a mild understatement. "The tube has to have a leak in it," he reasoned.

He repeated the experiment again—same result.

Then again—same result.

The fact is he made the water disappear again and again. Actually, what Cavendish had done was convert the liquid water to its gaseous state—into an invisible gas.

When Cavendish analyzed the contents of the tube, he found it contained a mixture of two gases, one of which was *inflammable air* and the other was a heavier gas. This heavier gas had only been discovered a few years earlier by his colleague Joseph Priestly (English chemist and clergyman) who, finding that it kept a mouse alive and supported combustion, called it *vital air*.

2.3.2 JUST TWO H's AND ONE O

Cavendish had been able to separate the two main constituents that make up water. All that remained was for him to put the ingredients back together again. He accomplished this by mixing a measured volume of inflammable air with different volumes of its vital counterpart, and setting fire to both. He found that most mixtures burned well enough, but when the proportions were precisely two to one, there was an explosion and the walls of his test tubes were covered with liquid droplets. He quickly identified these as water.

Cavendish made an announcement: Water was not water. Moreover, water is not just an odorless, colorless, and tasteless substance that lay beyond reach of chemical analysis. Water is not an element in its own right, but a compound of two independent elements, one a supporter of combustion and the other combustible. When united, these two elements become the preeminent quencher of thirst and flames.

It is interesting to note that a few years later, the great French genius, Antoine Lavoisier, tied the compound neatly together by renaming the ingredients *hydrogen*—"the water producer"—and *oxygen*. In a fitting tribute to his guillotined corpse (he was a victim of the revolution), his tombstone came to carry a simple and telling epitaph, a fitting tribute to the father of a new age in chemistry—*just two H's and one O*.

2.3.3 SOMEWHERE BETWEEN 0° AND 105°

We take water for granted now. Every high-school level student knows that water is a chemical compound of two simple and abundant elements.

And yet scientists continue to argue the merits of rival theories on the structure of water. The fact is we still know little about water. For example, we don't now know how water works.

Part of the problem lies with the fact that no one has ever seen a water molecule. It is true that we have theoretical diagrams and equations. We also have a disarmingly simple formula—H_2O. The reality, however, is that water is very complex. X-rays, for example, have shown that the atoms in water are intricately laced.

We have said over and over again that water is special, strange, and different. Water is also almost indestructible. Sure, we know that electrolysis can separate water atoms, but we also know that once they get together again they must be heated up to more than 2,900°C to separate them again.

Water is also idiosyncratic. This can be seen in the way in which the two atoms of hydrogen in a water molecule take up a very precise and strange (different) alignment to each other. Not at angles of 45°, 60°, or 90°—oh no, not water. Remember, water is different. The two hydrogen atoms *always* come to rest at an angle of approximately 105° from each other, making all diagrams of their attachment to the larger oxygen atom look like Mickey Mouse ears on a very round head.

This 105° relationship makes water lopsided, peculiar, and eccentric—it breaks all the rules. You're not surprised are you?

One thing is certain, however; this 105° angle is crucial to all life as we know it. Thus, the answer to defining why water is special, strange, different—and vital, lies somewhere between 0°–105°.

2.4 THE WATER CYCLE (HYDROLOGIC CYCLE)

Water is never stationary; it is constantly in motion. This phenomena occurs because of the water or *hydrologic cycle*. Because of the water cycle we can accurately say that the water created eons ago is still with us, still moving restlessly from state to state, and still flowing unimpaired down the passageways of time.

In simple terms the water cycle can be explained as follows: The sun helps transfer water from lakes and oceans to the land. As the sun shines on the earth, the surface water is heated and evaporates, forming an invisible gas that mixes with the air. This gas is *water vapor;* it is pure water without any minerals or bacteria in it. This water vapor rises in the air, then cools, and condenses into tiny drops of water that form clouds. Further cooling may form drops large enough to fall as rain. In this way the water is brought from the oceans to the land, where it reappears in springs and wells, soaks into the ground, or runs off again through streams and rivers back to the ocean. This phenomena is called the *water cycle.*

After having explained the water cycle in very simple terms to provide foundational information, it is important to point out that the actual movement of water on earth is much more complex. Three different methods of transport are involved in this water movement: *evaporation, precipitation,* and *run-off.*

Evaporation of water is a major factor in hydrologic systems. Evaporation is a function of temperature, wind velocity, and relative humidity. Evaporation (or vaporization) is, as the name implies, the formation of vapor. Dissolved constituents such as salts remain behind when water evaporates. Evaporation of the surface water of oceans provides most of the water vapor. It should be pointed out, however, that water can also vaporize through plants, especially from leaf surfaces. This process is called *evapotranspiration.* It may surprise you that ice can also vaporize without melting first. However, this *sublimation* process is slower than vaporization of liquid water.

Evaporation rates are measured with evaporation pans. These evaporation pans provide data that indicate the atmospheric evaporative demand of an area and can be used to estimate (1) the rates of evaporation from ponds, lakes, and reservoirs, and (2) evapotranspiration rates. It is important to note that several factors affect the rate of pan evaporation. These factors include the type of pan, type of pan environment, method of operating the pan, exchange of heat between pan and ground, solar radiation, air temperature, wind, and temperature of the water surface (Jones, 1992).

Precipitation includes all forms in which atmospheric moisture descends to earth—rain, snow, sleet, and hail. Before precipitation can occur, the water that enters the atmosphere by vaporization must first condense into liquid (clouds and rain) or solid (snow, sleet, and hail) before it can fall. This vaporization process absorbs energy. This energy is released in the form of heat when the water vapor condenses. You can best understand this phenomena when you compare it to what occurs when water that evaporates from your skin absorbs heat, making you feel cold. *Note:* The annual evaporation from ocean and land areas is the same as the annual precipitation.

Run-off is the flow back to the oceans of the precipitation that falls on land. This journey to the oceans is not always unobstructed—flow back may be intercepted by vegetation (from which it later evaporates), a portion is held in depressions, and a portion infiltrates into the ground. A part of the infiltrated water is taken up by plant life and returned to the atmosphere through evapotranspiration, while the remainder either moves through the ground or is held by capillary action. Eventually, water drips, seeps, and flows its way back into the oceans.

Assuming that the water in the oceans and ice caps and glaciers is fairly constant when averaged over a period of years, the water balance of the earth's surface can be expressed by the relationship: Water lost = Water gained (Turk & Turk, 1988).

2.5 WATER SOURCES

Where on planet earth is potable (drinking) water readily available for consumption? This question can be easily answered simply by referring to a map of the world. With such a map in hand, one only need look for those locations on the map that show population centers (cities). As a case in point, consider, for example, the United States. It was the presence of abundant water that attracted American settlers to establish settlements along rivers. The rivers provided the water settlers needed to sustain life, and the flow of the river water provided the principal source of power for early industry.

It is interesting to note that the earliest settlement of the United States occurred on its East Coast. In most cases (the early Jamestown, Virginia, settlement is an exception), settlers along this Eastern Seaboard area were lucky. They had settled along river systems of excellent quality. In addition, these river systems were ideally suited for paper and textile manufacturing.

Later, as settlers branched out inland (west) from their earlier settlements, they found that finding potable water was not all that easy to do. They found that the further west they travelled the salinity of the rivers and streams became higher. This was particularly the case with those rivers and streams that were long, flowing through and over areas of relatively soluble rock formations.

In Western regions of the United States (e.g., in deserts), the map of the United States shows sparse settlement. This is the case, of course, because of a lack of water; thus, these regions are occupied by fewer people and other species than other biomes.

You have heard the saying used in the real estate business that goes: Location! Location! Location! Location is everything. Well, in regards to the availability of potable water we can simply say that in the human settlement business the saying would be: Location is based on Water! Water! Water!

To sum all this up we can simply say that on land, the availability of a regular supply of potable water is the most important factor affecting the presence—or absence—of many lifeforms. Most people and other living things are found in regions of the United States and other parts of the world where potable water is readily available.

2.5.1 WHAT ARE THE MAJOR SOURCES OF DRINKING WATER?

Earlier it was pointed out that of the 326 million cubic miles of water covering earth only about 3% of this total is fresh, with most locked up in

TABLE 2.1. World Water Distribution.

Location	Percent of Total
Land areas	
Freshwater lakes	0.009
Saline lakes and inland seas	0.008
Rivers (average instantaneous volume)	0.0001
Soil moisture	0.005
Groundwater (above depth of 4000 m)	0.61
Ice caps and glaciers	2.14
	2.8
Atmosphere (water vapor)	0.001
Oceans	97.3
Total all locations (rounded)	100

Source: Adapted from Peavy et al. (1985), p. 12.

polar ice caps, in glaciers, in lakes, in flows through soil, and in river and stream systems. Out of this 3% only 0.027% is available for human consumption.

Exactly how is the world's water distributed? The data contained in Table 2.1 should help to answer this question.

After reviewing Table 2.1, it should be obvious that the major sources of drinking water are from surface water, groundwater, and from groundwater that is under the direct influence of surface water (i.e., a spring or a shallow well).

2.5.1.1 Surface Water

Simply stated, *surface water* is the water on the earth's surface as distinguished from subsurface water (groundwater). Most of this surface water is a product of precipitation in the form of rain, snow, sleet or hail. Surface water is exposed or open to the atmosphere and results from the movement of water on and just under the earth's surface (overland flow). Basically, what we are really saying here is that this overland flow is the same thing as surface runoff, which is the amount of rainfall which passes over the earth's surface. Specific sources of surface water include:

- rivers
- streams
- lakes
- impoundments (man-made lakes that are made by damming a river or stream)

- very shallow wells that receive input via precipitation
- springs that are affected by precipitation (i.e., their flow or quantity is directly affected by precipitation)
- rain catchments (drainage basin)
- tundra ponds or muskegs (peat bogs)

As a source of potable water, surface water has its advantages. For example, (1) surface water is usually easy to locate. The point is that you do not have to be a geologist or hydrologist to find it; and (2) normally, surface water is not tainted with chemicals that are precipitated from the earth's strata. As with just about everything else, when you point out the advantages of anything, you can also find disadvantages. This is also the case with surface water. The biggest disadvantage of using surface waters as a source of potable water is that they are easily contaminated (polluted) with microorganisms that can cause waterborne diseases and from chemicals that enter the surface waters from surrounding runoff and upstream discharges. There can also be problems with water rights.

In regards to water rights, if you are familiar with the battles that took place (some are still being waged) in the western United States during the early days when cattle folks were fighting homesteaders over rangeland, you are familiar with the cause: rights to surface water. Today, in most places, removal of water from a river, stream, spring or lake requires a legal permit.

Earlier it was pointed out that most surface water is the result of surface runoff. The amount and flow rate of this surface runoff is highly variable. This variability comes into play for two main reasons: (1) human interference (influences) and (2) natural conditions. In some cases, surface water runs quickly off land. This is generally undesirable from a water resources standpoint because it does not provide enough time for water to infiltrate into the ground and recharge groundwater aquifers. Another problem with surface water that runs quickly off land is erosion and flooding. Obviously, these two occurrences are unwanted. Probably the only good thing that can be said about surface water that quickly runs off land is that it does not have enough time (usually) to increase its mineral content. Surface water running slowly off land may be expected to have all the opposite effects.

Surface water runoff is influenced by several factors. However, before beginning a discussion of these factors, it is important to have some understanding of the movement of water.

The movement of surface water begins with its *drainage basin*. This surface water drainage basin is sometimes called drainage area, catchment, and/or watershed. When dealing with groundwater, this area is called the *recharge area* (i.e., the place from which precipitation flows into an underground water source). Whatever it is called, the drainage basin is the location from which surface water is obtained.

The surface water drainage basin is normally characterized as an area measured in square miles, acres, or sections. Obviously, if a city is taking water from a surface water source it is important to know the size of its drainage basin.

Surface water runoff, like the flow of electricity, flows or follows along the path of least resistance. Surface water within the drainage basin normally flows toward one primary watercourse (river, stream, brook, creek, etc.). This is the case unless some man-made distribution system (canal or pipeline) diverts the flow.

The question is: How rapidly does surface water run off the land surface? It depends.

There are factors that directly influence the surface water's flow over land. The principal factors are

- *rainfall duration:* The duration of a rainstorm influences the amount of runoff. This is the case because even a light, gentle rain, if it lasts long enough, can, with time, saturate the soil and allow runoff to take place.
- *rainfall intensity:* As rainfall increases in intensity, the surface of the soil quickly becomes saturated. This saturated soil can hold no more water; thus, as more rain falls and water builds up on the surface it begins to flow, creating surface runoff.
- *soil moisture:* Obviously, if soil is already wet or saturated from a previous rain, surface runoff will occur sooner than if the soil were dry. Thus, the amount of existing moisture in the soil has a definite impact on surface runoff. It should be pointed out that surface runoff from frozen soil can be up to 100% of snow melt or rain runoff because frozen ground is basically impervious.
- *soil composition:* The amount of runoff produced is directly related to the composition of the surface soil. Obviously, if the ground surface is of hard rock composition, 100% runoff will result. This is also the case with clay soils. Clay soils have very small void spaces that swell when wet. When the clay swells, the void spaces close and do not allow for infiltration. The opposite effect would result when the soil is made up of coarse sand. Coarse sand has large void spaces that allow easy flow of water through it. This is the case even if a torrential downpour occurs.
- *vegetation cover:* Runoff is limited by ground cover. Roots of vegetation and pine needles, pine cones, leaves, and branches create a porous layer (sheet of decaying natural organic substances) above the soil. This porous "organic" sheet (ground cover) readily allows water into the soil. Vegetation and organic waste also act as a cover to protect the soil from hard, driving rains. These hard, driving rains

can compact bare soils, close off void spaces, and increase run-off—vegetation and ground cover work to maintain the soil's infil-tration and water-holding capacity. It is also important to note that vegetation and ground cover also work to reduce evaporation of soil moisture.

- *ground slope:* Water flow off of flat land is usually so slow that there is time (opportunity) for a large portion of it to infiltrate the ground. However, this is not the case when rain falls on steeply sloping ground; up to 80+% of it may become surface runoff.
- *human influences:* Various human activities have a definite impact on surface water runoff. Most human activities tend to increase the rate of water flow. For example, canals and ditches are usually con-structed to provide steady flow, and agricultural activities generally remove ground cover that would work to retard the runoff rate. On the opposite extreme, man-made dams are generally built to retard the flow of runoff.

Paved streets, tarmac, paved parking lots, and buildings are impervious to water infiltration; they greatly increase the amount of runoff. Unfortu-nately, because of these impervious surfaces, which work to hasten the flow of surface water, flooding often occurs with devastating conse-quences. Moreover, because of the increased runoff caused by increased urbanization, a whole new industry (field) has developed: Storm Water Management. Along with storm water management problems, there is an-other problem with runoff on impervious surfaces. Namely, water does not have the opportunity to infiltrate (percolate) into the soil to replenish (re-charge) groundwater supplies. Obviously, this is a result that can have dev-astating impact on a location's water supply.

2.5.1.2 Groundwater

Approximately three feet of water falls each year on every square foot of land. About six inches of this goes back to the sea. Evaporation takes up about two feet. What remains, approximately six inches, seeps into the ground, entering and filling every interstice, each hollow and cavity, like an absorbent. Although comprised of only 1/6 of the total (1,680,000 miles of it), if it could be ladled up and spread out over the earth's surface, it would blanket all land to a depth of 1,000 feet.

This gigantic water source (literally an ocean beneath our feet) forms a reservoir that feeds all the natural fountains and springs of earth. Eventually, it works its way to the surface. Some comes out clean and cool, a liquid blue-green phantom; and some, occupying the deepest recesses, pressurizes and shoots back to the surface in white, foamy, wet chaos, as geysers.

Fortunately, most of the rest lies within easy reach, just beneath the surface. This is groundwater.

Water falling to the ground as precipitation follows three courses. Some runs off directly to rivers and streams, some infiltrates to ground reservoirs, and the rest evaporates or transpires through vegetation. The water in the ground (groundwater) is "invisible" and may be thought of as a temporary natural reservoir (ASTM, 1969). Almost all groundwater is in constant motion toward rivers or other surface water bodies. In fact, most of the water discharging through streams to the oceans includes most of the runoff from groundwater reservoirs.

Groundwater is defined as water that is below the earth's crust, but not more than 2,500 feet below the crust. Thus, if water is located between the earth's crust and the 2,500 foot level it is considered usable (potable) fresh water (Arasmith, 1993). In the United States, it is estimated "that at least 50% of total available fresh water storage is in underground aquifers" (Kemmer, 1979, p. 17).

Groundwater is usually obtained from springs that are not influenced by surface water, a local hydrologic event, or from wells.

Groundwater, in relationship to surface water, has several advantages: (1) unlike surface water, groundwater is not easily contaminated; (2) groundwater sources are usually lower in bacteriological contamination than surface waters; (3) the quality of groundwater usually remains stable throughout the year; and (4) in the United States, for example, groundwater is available in most locations.

When comparing groundwater with surface water sources, there are some disadvantages in using groundwater: (1) if there is contamination, it is often hidden from view; (2) groundwater is usually loaded with minerals (increased level of hardness) because it is in contact longer with minerals; (3) when groundwater supplies are contaminated it is very difficult to remove the contaminants; (4) because it must be pumped from the ground, operating costs are usually higher; and (5) if the groundwater source is located near coastal areas, it may be subject to salt water intrusion.

2.6 WATER USE

In the United States, rainfall averages approximately $4,250 \times 10^9$ gallons a day. About two thirds of this returns to the atmosphere through evaporation directly from the surface of rivers, streams, and lakes and transpiration from plant foliage. This leaves approximately $1,250 \times 10^9$ gallons a day to flow across or through the earth to the sea (Kemmer, 1979).

The question is: Of the billions of gallons of water that are available for use in the United States, where is this water used? The National Academy of Sciences (1962), for example, estimates that approximately 310 billion gallons per day (bgd) are withdrawn; 142 bgd are used for irrigation; 142 bgd are used for industry (principally utility cooling water—100 bgd); 26 bgd are used in municipal application; 90 bgd are consumed (principally irrigation loss to ground and evaporation); and 220 bgd are returned to streams.

In this text we are primarily concerned with water use in regards to municipal applications (demand). Municipal water demand is usually classified according to the nature of the user. These classifications are:

(1) *Domestic:* Domestic water is supplied to houses, schools, hospitals, hotels, restaurants, etc. for culinary, sanitary, and other purposes. Use varies with the economic level of the consumer, the range being 20 to 100 gallons per capita per day. It should be pointed out that these figures include water used for watering gardens, lawns and washing cars.

(2) *Commercial and industrial:* Commercial and industrial water is supplied to stores, offices, and factories. The importance of commercial and industrial demand is based, of course, on whether there are large industries that use water supplied from the municipal system. These large industries demand a quantity of water directly related to the number of persons employed, to the actual floor space or area of each establishment, and to the number of units manufactured or produced. Arasmith (1993) estimates that industry in the United States uses an average of 150 bgd of water each day.

(3) *Public use:* Public use water is the water furnished to public buildings and used for public services. This includes water for schools, public buildings, fire protection, and for flushing streets.

(4) *Loss and waste:* Water that is lost or wasted (i.e., unaccounted for) is attributable to leaks in the distribution system, inaccurate meter readings, and for unauthorized connections. Loss and waste of water can be expensive. In order to reduce loss and waste a regular program that includes maintenance of the system and replacement and/or recalibration of meters is required (McGhee, 1991).

Water use is covered in greater detail in Chapter 10 of this text.

CHAPTER 3

Water Mathematics

Those who have difficulty in math often do not lack the ability for mathematical calculation, they merely have not learned, or have not been taught, the "language of math." (Price, 1991, p. vii)

3.1 INTRODUCTION

A water/wastewater (w/ww) operator and other water specialists must have a high level of common sense, good judgement, and sufficient knowledge of the basics of mathematics, hydraulics, chemistry, ecology, biology, and other related subjects. The w/ww operator also needs extensive training and experience in the operation, maintenance, replacement, and repair of w/ww treatment plant equipment. The point is the professional w/ww operator must know more than just how to operate his/her plant—he/she must understand the "how and why" of treatment processes.

The reader may be wondering: What exactly are the duties of the w/ww operator? Good question.

Before explaining exactly what is expected of w/ww operators, it is important to state again that these highly skilled professionals must know more than just how to operate the plant—they must be "generalists" with a wide range of technical knowledge. When you think about it, it just makes good common sense for w/ww operators to be armed with extensive knowledge and experience. When fully armed, the operator is not only better able to perform his/her job in a highly professional manner but also has the knowledge and experience to protect the consumer and the environment—and to qualify for and pass state licensing examinations.

Water and wastewater operators are expected to

- operate mechanical equipment, including bar screens, motors, pumps, conveyors, filters, incinerators, compost facilities, chemical feeders, and meters

TABLE 3.1. Mathematical Computations Routinely Made by W/WW Operators (a Representative Sample).

Operation	Computations Routinely Made to Determine
Water treatment	Per capita water use
	Domestic water use based on fixed rates
	Water use per unit of industrial product produced
	Average daily flow
	Surface overflow rate
	Weir overflow rate
	Filter loading rate
	Filter backwash rate
	Mudball calculations
	Detention time
	Well problems
Wastewater treatment	Tank volume
	Channel or pipeline volume
	Flow and velocity rates and calculations
	Average flow rates and conversions
	Chemical dosage calculations
	Loading calculations for BOD-COD and SS
	BOD and SS removal
	WAS pumping rates
	Hydraulic loading rate
	Filtration rate
	Surface overflow rate
	Backwash rate
	Organic rate
	Food/microorganism ratio
	Digester loading rate
	Digester volatile solids loading
	MCRT
	Unit process efficiency
	Density and specific gravity
	Horsepower
	Pump capacity
	Wet well capacity
	Hydraulic loading rates

- operate electrical and electronic equipment, including motor controllers, recorders, automatic monitors, and emergency/backup power systems
- maintain, service, repair, replace, and calibrate various mechanical, electrical, and electronic equipment
- determine proper chemical dosages and control chemical applications for the treatment processes
- inventory, order, and store chemicals and spare parts for equipment
- maintain a safe working environment

- collect samples for testing
- keep accurate and complete records of treatment operation and submit required reports to regulatory agencies
- perform laboratory analyses
- perform regular, routine preventive maintenance on various types of equipment
- keep informed on regulatory requirements affecting the water/ wastewater system
- perform general plant housekeeping and maintenance
- recommend to supervisors any repairs, improvements, or replacements that should be made to the treatment system
- keep current (informed) on new developments in technology

After reviewing the above list, it should be obvious that in order for the w/ww treatment plant operator to perform these duties successfully, he/she must have a great deal of knowledge and experience. Simply stated: The water/wastewater operator must be a "Jack or Jill" of all trades (Spellman, 1996a).

At this point you might be willing to buy the argument that w/ww operators and water specialists must be well-trained and experienced to perform their duties in a highly professional manner. However, you might also ask: Why is mathematics so important? Another good question.

Answer: Water and wastewater treatment plant operations involve a large number of process control calculations. All of these calculations are based on principles of basic mathematics.

Further, to demonstrate a few of the process control calculations required to be made by w/ww operators, Table 3.1 is provided.

3.2 BASIC MATH

Note: It is assumed that water/wastewater operators/and other water specialists have a fundamental knowledge of basic mathematical operations. Thus, the purpose of the following sections is to provide only a brief review of the mathematical concepts and applications frequently employed in actual process control activities.

3.3 CONVERSIONS OF FRACTIONS TO DECIMALS

A fraction is an incomplete division. With the availability of calculators, the easiest method for working with fractions is to convert the fraction into decimal form. To convert the fraction, divide the numerator (top number) by the denominator (bottom number).

$$\text{Fraction} = \frac{\text{Numerator}}{\text{Denominator}}$$

Let's try a fraction to decimal conversion, What is the decimal equivalent of the fraction 5/8?

$$5/8 = 5 \div 8 = .0625$$

3.4 POWERS OF TEN AND SCIENTIFIC NOTATION

Note: In practice, the water/wastewater operator and water specialist should realize that the accuracy of a final answer can never be better than the accuracy of the data used. Having stated the obvious, it should also be pointed out that correct and accurate data is worthless unless the operator is able to make correct computations.

Two common methods of expressing a number—*powers of ten* and *scientific notation*—will be discussed in this section.

3.4.1 POWERS OF TEN NOTATION

An expression such as 4^7 is a shorthand method of writing multiplication. For example, 4^7 can be written as

$$4 \times 4 \times 4 \times 4 \times 4 \times 4 \times 4$$

The expression 4^7 is referred to as *4 to the seventh power* and is composed of an *exponent* and a *base* number. An exponent (or power of) indicates how many times a member is to be multiplied together. The base is the number being multiplied.

$$5^{7 \text{ (exponent)}}$$
$$\uparrow_{\text{(base)}}$$

Let's look at base 5 again, using different exponents.

This is referred to as *5 to the second power*, or *5 squared*. In expanded form,

$$5^2 = (5)(5)$$

The expression *5 to the third power* (called 5 cubed) is written as

$$5^3$$

In expanded form, this notation means

$$5^3 = (5)(5)(5)$$

These same considerations apply to letters (a, b, x, z, etc.) as well. For example:

$$z^2 - (z)(z) \text{ or } z^4 = (z)(z)(z)(z)$$

When a number or letter does not have an exponent it is considered to have an exponent of one.

$$\text{Thus } 5 = 5^1 \text{ or } z = z^1$$

The following examples help to illustrate the concept of powers notation.

Example 1

How is the term 3^3 written in expanded form? The power (exponent) of 3 means that the base number is multiplied by itself three times:

$$3^3 = (3)(3)(3)$$

Example 2

How is the term in.^{2} written in expanded form? The power (exponent) of 2 means that the term is multiplied by itself two times:

$$\text{in.}^2 = (\text{in.})(\text{in.})$$

Example 3

How is the term 7^5 written in expanded form? The exponent of 5 indicates that 7 is multiplied by itself five times:

$$7^5 = (7)(7)(7)(7)(7)$$

Example 4

How is the term x^6 written in expanded form? The exponent of 6 indicates that x is multiplied by itself six times:

$$x^6 = (x)(x)(x)(x)(x)(x)$$

Example 5

How is the term $(3/4)^2$ written in expanded form? When parentheses are used, the exponent refers to the entire term within the parentheses. Thus, in this example, $(3/4)^2$ means

$$(3/4)^2 = (3/4)(3/4)$$

When a negative exponent is used with a number or term, a number can be reexpressed using a positive exponent:

$$5^{-3} = 1/5^3$$

Another example is

$$10^{-5} = 1/10^5$$

Example 6

How is the term 9^{-3} written in expanded form?

$$9^{-3} = \frac{1}{9^3} = \frac{1}{(9)(9)(9)}$$

Note: Any number or letter such as 3^0 or X^0 does not equal 3×1 or $X1$, but simply 1.

Example 7

When a term is given in expanded form, you can determine how it would be written in exponential form. For example,

$$(5)(5)(5) = 5^3$$

or

$$(in.)(in.) = in.^2$$

Example 8

Write the following term in exponential form:

$$\frac{(3)(3)}{(7)(7)(7)}$$

The exponent for the numerator [remember: (numerator/denominator)] of the fraction is 2 and the exponent for the denominator is 3. Therefore, the term would be written as

$$\frac{(3)^2}{(7)^3}$$

It should be pointed out that since the exponents are not the same, parentheses cannot be placed around the fraction and a single exponent cannot be used.

Example 9

It is common to see powers used with a number or term used to denote area or volume units (in.^{2}, ft^2, in.^{3}, ft^3) and in scientific notation.
Write the following term in exponential form:

$$\frac{(\text{in.})\,(\text{in.})}{(\text{ft})\,(\text{ft})}$$

The exponent of both the numerator and denominator is 2:

$$\frac{(\text{in.})^2}{(\text{ft})^2}$$

The exponents are the same which allows the use of parentheses to express the term as follows:

$$\left(\frac{\text{in.}}{\text{ft}}\right)^2$$

Example 10

In moving a power from the numerator of a fraction to the denominator, or vice versa, the sign of the exponent is changed. For example

$$\frac{3^3 \times 4^{-2}}{8} = \frac{3^3}{8 \times 4^2}$$

3.4.2 SCIENTIFIC NOTATION

Scientific notation is a method by which any number can be expressed as a term multiplied by a power of ten. The term is always greater than or equal to 2 but less than 10. Examples of powers of ten are

$$3.2 \times 10^1$$
$$1.7 \times 10^3$$
$$9.550 \times 10^4$$
$$3.51 \times 10^{-2}$$

The numbers can be taken out of scientific notation by performing the indicated multiplication. For example,

$$3.2 \times 10^1 = (3.2)(10)$$
$$= 32$$
$$1.7 \times 10^3 = (1.7)(10)(10)(10)$$
$$= 1,700$$
$$9.550 \times 10^4 = (9.550)(10)(10)(10)(10)$$
$$= 95,500$$
$$3.51 \times 10^{-2} = (3.51)1/10^2$$
$$= (3.51)\frac{(1)}{(10)(10)}$$
$$= 0.0351$$

An easier way to take a number out of scientific notation is by moving the decimal point the number of places indicated by the exponent.

RULE 1

Multiply by the power of ten indicated. A positive exponent indicates a decimal move to the *right,* and a negative exponent indicates a decimal move to the *left.*

Example 11

Using the same examples above, the decimal point move rather than the multiplication method is performed as follows:

$$3.2 \times 10^1$$

The positive exponent of 1 indicates that the decimal point in 3.2 should be moved one place to the right:

$$3.2 \times 32$$

The next example is

$$1.7 \times 10^3$$

The positive exponent of 3 indicates that the decimal point in 1.7 should be moved three places to the right:

$$1.700 \times 1,700$$

The next example is

$$9.550 \times 10^4$$

The positive exponent of 4 indicates that the decimal point should be moved four places to the right:

$$9.5500 = 95,500$$

The final example is

$$3.51 \times 10^{-2}$$

The negative exponent of 2 indicates that the decimal point should be moved two places to the left:

$$03.51 \times 0.351$$

Example 12

Take the following number out of scientific notation:

$$3.516 \times 10^4$$

The positive exponent of 4 indicates that the decimal point should be moved 4 places to the right.

$$3.5160 = 35,160$$

Example 13

Take the following number out of scientific notation:

$$3.115 \times 10^{-4}$$

$$.0003.115 = 0.0003115$$

There are very few instances in which you will need to put a number or numbers *into* scientific notation, but you should know how to do it, if required. Thus, the method is discussed below. However, before demonstrating the process of putting a number into scientific notation, it important to point out the procedure and the rule involved with the process.

Procedure: When placing a number *into* scientific notation, place a decimal point after the first nonzero digit. (Remember that if no decimal point is shown in the number to be converted, it is assumed to be at the end of the number). Count the number of places from the standard position to the original decimal point. This represents the exponent of the power of ten.

RULE 2

When a number is put *into* scientific notation, a decimal point move to the *left* indicates a *positive* exponent; a decimal point move to the *right* indicates a *negative* exponent.

Now let's try converting a few numbers into scientific notation.

First, convert 69 into scientific notation.

Note: In order to obtain a number between 1 and 9, the decimal point must be moved one place to the left. This move of one place gives the exponent, and the move to the left means that the exponent is positive:

$$69 = 6.9 \times 10^1$$

Let's try converting another number

$$1,400$$

Remember, in order to obtain a number between 1 and 9, the decimal point must be moved three places to the left. The number of place moves (3) becomes the exponent of the power 10, and the move to the left indicates a positive exponent:

$$1,400 = 1.4 \times 10^3$$

Let's try a decimal number

$$0.0561$$

$$0.0561 = 5.61 \times 10^{-2}$$

Example 14

Put the following number into scientific notation:

$$5,115,000$$

$$5,115,000 = 5.115 \times 10^{6}$$

Example 15

Convert the following decimal to scientific notation:

$$0.000525$$

$$0.000525 = 5.25 \times 10^{-4}$$

3.5 DIMENSIONAL ANALYSIS

Dimensional analysis is a valuable tool used as a way to check if you have set up a problem correctly. In using dimensional analysis to check a math setup, you work with the dimensions (units of measure) only—not with the numbers. In order to use the dimensional analysis method, you must know how to perform three basic operations:

Basic Operation:

(1) To complete a division of units, always ensure that all units are written in the same format; it is best to express a horizontal fraction (such as gal/ft³) as a vertical fraction.

horizontal to vertical

$$\text{gal/cu ft to } \frac{\text{gal}}{\text{cu ft}}$$

$$\text{psi to } \frac{\text{lb}}{\text{sq in.}}$$

Let's apply these procedures in the following examples shown below.

$$\text{ft}^3/\text{min becomes} \quad \frac{\text{ft}^3}{\text{min}}$$

$$\text{s}/\text{min becomes} \quad \frac{\text{s}}{\text{min}}$$

Basic Operation:

(2) You must know how to divide by a fraction. Let's try dividing by a fraction. For example,

$$\frac{\dfrac{\text{lb}}{\text{d}}}{\dfrac{\text{min}}{\text{d}}} \quad \text{becomes} \quad \frac{\text{lb}}{\text{d}} \times \frac{\text{d}}{\text{min}}$$

In the above problem, you may have noticed that the terms in the denominator were inverted before the fractions were multiplied. This is a standard rule that must be followed when dividing fractions. Another example is

$$\frac{\dfrac{\text{mm}^2}{\text{mm}^2}}{\dfrac{}{\text{m}^2}} \quad \text{becomes} \quad \text{mm}^2 \times \frac{\text{m}^2}{\text{mm}^2}$$

Basic Operation:

(3) You must know how to cancel or divide terms in the numerator and denominator of a fraction.

After fractions have been rewritten in the vertical form and division by the fraction has been reexpressed as multiplication as shown above, then the terms can be canceled (or divided) out.

Note: For every term that is canceled in the numerator of a fraction, a similar term must be canceled in the denominator, and vice versa, as shown below:

$$\frac{\text{kg}}{\cancel{\text{d}}} \times \frac{\cancel{\text{d}}}{\text{min}} = \frac{\text{kg}}{\text{min}}$$

$$\cancel{mm}^2 \times \frac{m^2}{\cancel{mm}^2} = m^2$$

$$\frac{\cancel{gal}}{min} \times \frac{ft^3}{\cancel{gal}} = \frac{ft^3}{min}$$

QUESTION: How are units that include exponents calculated?

When written with exponents, such as ft³, a unit can be left as is or put in expanded form, (ft)(ft)(ft), depending on other units in the calculation. The point is that it is important to ensure that square and cubic terms are expressed uniformly, as sq ft, cu ft, or as ft², ft³. For dimensional analysis, the latter system is preferred.

For example, let's say that you wish to convert 1,400 ft³ volume to gallons, and you will use 7.48 gal/ft³ in the conversion. The question becomes: Do you multiply or divide by 7.48?

In the above instance, it is possible to use dimensional analysis to answer this question; that is, are we to multiply or divide by 7.48?

In order to determine if the math setup is correct, *only the dimensions* are used.

First, try dividing the dimensions:

$$\frac{ft^3}{gal/ft^3} = \frac{ft^3}{\dfrac{gal}{ft^3}}$$

Then the numerator and denominators are multiplied to get

$$= \frac{ft^6}{gal}$$

So, by dimensional analysis you determine that if you divide the two dimensions (ft³ and gal/ft³), the units of the answer are ft⁶/gal, *not* gal. It is clear that division is not the right way to go in making this conversion.

What would have happened if you had multiplied the dimensions instead of dividing?

$$(ft^3)(gal/ft^3) = (ft^3)\left(\frac{gal}{(ft^3)}\right)$$

Then multiply the numerator and denominator to obtain

$$= \frac{(ft^3)(gal)}{ft^3}$$

And cancel common terms to obtain

$$= \frac{(\cancel{ft^3})(gal)}{\cancel{ft^3}}$$

$$= gal$$

Obviously, by multiplying the two dimensions (ft³ and gal/ft³), the answer will be in gallons, which is what you want. Thus, since the math setup is correct, you would then multiply the numbers to obtain the number of gallons.

$$(1,400 \text{ ft}^3)(7.48 \text{ gal/ft}^3) = 10,472 \text{ gal}$$

Now let's try another problem with exponents. You wish to obtain an answer in square feet. If you are given the two terms—70 ft³/s and 4.5 ft/s—is the following math setup correct?

$$(70 \text{ ft}^3/s)(4.5 \text{ ft/s})$$

First, only the dimensions are used to determine if the math setup is correct. By multiplying the two dimensions, you get

$$(ft^3/s)(ft/s) = \left(\frac{ft^3}{s}\right)\left(\frac{ft}{s}\right)$$

Then multiply the term in the numerators and denominators of the fraction:

$$= \frac{(ft^3)(ft)}{(s)(s)}$$

$$= \frac{ft^4}{s^2}$$

Obviously, the math setup is incorrect because the dimensions of the answer are not square feet. Therefore, if you multiply the numbers as shown above, the answer will be wrong.

Let's try division of the two dimension instead.

$$\text{ft}^3/\text{s} = \dfrac{\dfrac{\text{ft}^3}{\text{s}}}{\dfrac{\text{ft}}{\text{s}}}$$

Invert the denominator and multiply to get

$$= \left(\dfrac{\text{ft}^3}{(\text{s})}\right)\left(\dfrac{\text{s}}{(\text{ft})}\right)$$

$$= \dfrac{(\text{ft})(\text{ft})(\text{ft})(\text{s})}{(\text{s})(\text{ft})}$$

$$= \dfrac{(\cancel{\text{ft}})(\text{ft})(\text{ft})(\cancel{\text{s}})}{(\cancel{\text{s}})(\cancel{\text{ft}})}$$

$$= \text{ft}^2$$

Since the dimensions of the answer are square feet, this math setup is correct. Therefore, by dividing the numbers as was done with units, the answer will also be correct.

$$\dfrac{70 \text{ ft}^3/\text{s}}{4.5 \text{ ft}/\text{s}} = 15.56 \text{ ft}^2$$

Example 16

You are given two terms—5 m/s and 7 m²—and the answer to be obtained is in cubic meters per second (m³/s). Is multiplying the two terms the correct math setup?

$$(\text{m}/\text{s})(\text{m}^2) = \dfrac{\text{m}}{\text{s}} \times \text{m}^2$$

Multiply the numerators and denominator of the fraction:

$$= \frac{(m)(m^2)}{s}$$

$$= \frac{m^3}{s}$$

Since the dimensions of the answer are cubic meters per second (m³/s), the math setup is correct. Therefore, multiply the numbers to get the correct answer.

$$5 \ (m/s)(7 \ m^2) = 35 \ m^3/s$$

Example 17

Solve the following problem:
Given: The flow rate in a water line is 2.1 ft³/s. What is the flow rate expressed as gallons per minute?
Set up the math problem and then use dimensional analysis to check the math setup:

$$(2.1 \ ft^3/s)(7.48 \ gal/ft^3)(60 \ s/min)$$

Dimensional analysis is used to check the math setup:

$$(ft^3/s)(gal/ft^3)(s/min) = \left(\frac{(ft^3)}{s}\right)\left(\frac{gal}{ft^3}\right)\left(\frac{s}{min}\right)$$

$$= \left(\frac{ft^3}{s}\right)\left(\frac{gal}{ft^3}\right)\left(\frac{s}{min}\right)$$

$$= \frac{gal}{min}$$

The math setup is correct as shown above. Therefore, this problem can be multiplied out to get the answer in correct units.

$$(2.1 \ ft^3/s)(7.48 \ gal/ft^3)(60 \ s/min) = 942.48 \ gal/min$$

3.6 ROUNDING OFF A NUMBER

It is sometimes necessary to round off measurements to a certain number of significant figures. The number of significant figures in a measurement is the number of digits that are known for sure, plus one more digit that is an estimate. The number 5.85436 cm, for example, has six significant figures. For numbers less than one, such as 0.00222, the zeros to the right of the decimal point are not considered significant figures; thus, 0.00222 has three significant figures. For numbers greater than ten, such as 222,000, the zeros should not be considered significant; thus the number 222,000 has three significant figures.

Rounding off a number means replacing the final digit of a number with zeros, thus expressing the number as tens, hundreds, thousands, ten thousands, or tenths, hundredths, thousandths, ten thousandths, etc. (e.g., 398 as 400; 0.39 as 0.4; or 6,669,969 as 7,000,000).

There is a basic rule to be followed when rounding off numbers.

RULE 3

A number is rounded off by dropping one or more numbers from the right, and adding zeros if necessary to place the decimal point. If the last figure dropped is 5 or more, increase the last retained figure by 1. If the last figure dropped is less than 5, do not increase the last retained figure.

Example 18: Rounding to significant figures

Round off 11,547 to 4, 3, 2, and 1 significant figure.

Solution

$$11,547 = 11,550 \text{ to 4 significant figures}$$
$$11,547 = 11,500 \text{ to 3 significant figures}$$
$$11,547 = 11,000 \text{ to 2 significant figures}$$
$$11,547 = 10,000 \text{ to 1 significant figure}$$

Example 19: Rounding to a particular place value in the decimal system

Round 47,936 to the nearest hundred.

The procedure used in this rounding depends on the digit just to the right of the hundreds place:

$$47,936$$
↑
(hundreds place)

Since the digit to the right of the hundreds place is less than 5, 9 is not changed and all the digits to the right of 9 are replaced with zeros: 47,936 becomes 47,900 (rounded to the nearest hundred).

Let's look at another example where a decimal number is to be rounded. Round 6.654 to the nearest tenth.

$$6.654$$
↑
(tenths place)

The digit to the right of the tenths place is 5. Therefore the 6 is increased by 1 and all digits to the right are dropped: 6.654 becomes 6.7 (rounded to the nearest tenth).

3.7 EQUATIONS: SOLVING FOR THE UNKNOWN

In water/wastewater treatment plant operations, you may use equations for various calculations. To make these calculations, you must first know the values for all but one of the terms of the equation to be used.

An *equation* is a statement that two expressions or quantities are equal in value. The statement of equality $5x + 4 = 19$ is an equation; that is, it is algebraic shorthand for "The sum of 5 times a number plus 4 is equal to 19." It can be seen that the equation $5x + 4 = 19$ is much easier to work with than the equivalent sentence.

When thinking about equations, it is helpful to consider an equation as being similar to a balance. The equal sign tells you that two quantities are "in balance" (i.e., they are equal).

Let's get back to the equation $5x + 4 = 19$. The solution to this problem may be summarized in three steps.

(1) $5x + 4 = 19$
(2) $5x = 15$
(3) $x = 3$

Step 1 expresses the whole equation. In Step 2, 4 has been subtracted from both members of the equation. In Step 3, both members have been divided by 5.

An equation is therefore kept in balance (both sides are kept equal) by subtracting the same number from both members (sides), adding the same number to both, or dividing or multiplying both by the same number.

The expression $5x + 4 = 19$ is called a *conditional equation* because it is true only when x has a certain value. The number to be found in a conditional equation is called the *unknown number;* the *unknown quantity;* or, more briefly, the *unknown.*

The point is that *solving an equation* is finding the value or values of the unknown that make the equation true.

Another equation, one that water/wastewater operators should be familiar with, is

$$Q = AV$$

where

Q = flow rate
A = area
V = velocity

The terms of the equation are Q, A, and V. In solving problems using this equation, you would need to be given values to substitute for any two of the three terms. Again, the term for which you do not have information is called the unknown. This unknown value is often indicated by a letter such as x, y, or z, but may be any other letter such as Q, A, V, etc.

Suppose you have this equation:

$$60 = (x)(5)$$

How can you determine the value of x? By following the axioms presented in this section, the solution to the unknown is quite simple.

It is important to point out that the following discussion includes only what the axioms are and how they work. (If you are interested in *why* these axioms work and how they came about, you should consult an algebra text.)

3.7.1 AXIOMS

(1) If equal numbers are added to equal numbers, the sums are equal.
(2) If equal numbers are subtracted from equal numbers, the remainders are equal.
(3) If equal numbers are multiplied by equal numbers, the products are equal.

(4) If equal numbers are divided by equal numbers (except zero), the quotients are equal.
(5) Numbers that are equal to the same number or to equal numbers are equal to each other.
(6) Like power of equal numbers are equal.
(7) Like roots of equal numbers are equal.
(8) The whole of anything equals the sum of all its parts.

Note: Axioms 2 and 4 were used to solve the equation $5x + 4 = 19$.

3.7.2 SOLVING SIMPLE EQUATIONS

As stated earlier, solving an equation is determining the value or values of the unknown number or numbers in that equation.

Example 20

Find the value of x if $x - 6 = 3$.

Here you can see by inspection that $x = 9$, but inspection does not help in solving more complicated equations. But if you notice that to determine that $x = 9$, 6 is added to each member of the given equation, you have acquired a method or procedure that can be applied to similar but more complex problems.

Solution   Given equation:

$$x - 6 = 3$$

Add 6 to each member (axiom 1),

$$x = 3 + 6$$

Collecting the terms (that is, adding 3 and 6),

$$x = 9$$

Example 21

Solve for x, if $3x - 5 = 7$ (each side is in simplest terms)

Solution $3x = 7 + 5$ [the term (-5) is moved to the right of the equal sign as $(+5)$]

$3x = 12$ (pure numbers are combined)

$$\frac{3x}{3} = \frac{12}{3}$$ (divide both sides)

$x = 4$ (we're done; x is alone on the left and is equal to the value on the right)

Example 22

Solve for x, if $x + 10 = 13$

Solution   Given equation:

$$x + 10 = 13$$

Subtracting 10 from each member (axiom 2),

$$x = 13 - 10$$

Collecting the terms,

$$x = 3$$

Example 23

Solve for x, if $5x + 5 - 7 = 3x + 6$

Solution   Given equation:

$$5x + 5 - 7 = 3x + 6$$

Collect the terms (+5) and (−7):

$$5x - 2 = 3x + 6$$

Add 2 to both members (axiom 1):

$$5x = 3x + 8$$

Subtract $3x$ from both members (axiom 2)

$$2x = 8$$

Divide both members by 2 (axiom 4):

$$x = 4$$

3.7.3 CHECKING THE ANSWER

After you have obtained a solution to an equation, you should always check it. This is an easy process. All you need do is substitute the solution for the unknown quantity in the given equation. If the two members of the equation are then identical, the number substituted is the correct answer.

Example 24

Solve and check $4x + 5 - 7 = 2x + 6$

Solution

$$4x + 5 - 7 = 2x + 6$$
$$4x - 2 = 2x + 6$$
$$4x = 2x + 8$$
$$2x = 8$$
$$x = 4$$

Substituting the answer $x = 4$ in the original equation,

$$4x + 5 - 7 = 2x + 6$$
$$4(4) + 5 - 7 = 2(4) + 6$$
$$16 + 5 - 7 = 8 + 6$$
$$14 = 14$$

Because the statement $14 = 14$ is true, the answer $x = 4$ must be correct.

3.7.4 SETTING UP EQUATIONS

The equations discussed in the preceding paragraphs were expressed in *algebraic* language. It is important to learn how to set up an equation by translating a sentence into an equation (into algebraic language) and then solve this equation. The following suggestions and examples should help you:

(1) Always read the statement of the problem carefully.
(2) Select the unknown number and represent it by some letter. If more than one unknown quantity exists in the problem, try to represent those numbers in terms of the same letter—that is, in terms of one quantity.
(3) Develop the equation, using the letter or letters selected, and then solve.

Example 25

Given: 2 less than 3 times a number is the same as 5 more than twice the number. Find the number.

Solution   Let n represent the number.

$$3n - 2 = 2n + 5$$
$$3n - 2n = 5 + 2$$
$$n = 7$$

The number is 7.

Example 26

Given: If 13 is decreased by the sum of a number and 2, the result is 4 less than the number. Find the number.

Solution   Let n represent the number.

$$13 - (n + 2) = n - 4$$
$$13 - n - 2 = n - 4$$
$$11 - n = n - 4$$
$$-n - n = -4 - 11$$
$$-2n = -15$$
$$\frac{-2n}{-2} = \frac{-15}{-2}$$
$$n = 7.5$$

The number is 7.5.

Example 27

Given: The greater of two numbers is 3 less than 7 times the smaller. Also, 12 more than the greater is the same as 10 times the smaller. Find both numbers.

Solution   Let n represent the small number.
Then $7n - 3$ must represent the large number.

$$7n - 3 + 12 = 10n$$
$$7n + 9 = 10n$$
$$7n - 10n = -9$$
$$-3n = -9$$
$$\frac{-3n}{-3} = \frac{-9}{-3}$$
$$n = 3$$

The small number is 3.

$$7(n) - 3$$

$$7(3) - 3 = 18$$

The large number is 18.

3.8 RATIO AND PROPORTION

3.8.1 RATIO

Ratio is the comparison of two numbers by division or an indicated division. The ratio of one number to another is determined when the one number is divided by the other.

A ratio always includes two numbers. For example: If a box has a length and width of 8 in. and 4 in., the ratio of the *length to the width* is expressed as 8/4 or 8:4. Both expressions have the same meaning.

All ratios are reduced to the lowest possible terms. This is similar to reducing a fraction to the lowest possible terms. The ratio of 8/4 or 8:4 should be reduced to its lowest possible terms by dividing the 8 and the 4 by 4. The resulting ratio is 2:1 or 2/1.

The ratio of the length of the box to its width is 2:1, since the box is 2 times as long as it is wide.

This ratio can also be stated as the relationship of the *width to the length*. The box is 4 in. wide and 8 in. long. The ratio of the width to the length is 4:8 or 4/8. This ratio when reduced becomes 1:2 or 1/2. The width of the box is 1/2 its length.

3.8.2 PROPORTIONS

Simply put, a *proportion* is a statement of equality between two ratios. Thus 2:4 = 4:8 is a proportion. It is evident that the two ratios are equal when the proportion is written in fractional form: 2/4 = 4/8. Either form may be read as follows: "Two is to four as four is to eight."

A general statement of the preceding proportion would be:

$$\frac{a}{b} = \frac{c}{d} \text{ (fractional form)}$$

$$a:b = c:d \text{ (proportional form)}$$

where *a, b, c,* and *d* represent numbers.

A proportion may be written with a double colon (::) in place of the equal sign:

$$a:b::c:d$$

The first and last terms of a proportion are called the *extremes;* the second and third terms are called the *means.* Thus in the preceding examples the *a* and *d,* and the 2 and 6, are the extremes; the *b* and *c,* and the 3 and 4, are the means.

By consideration of proportions as fractions, it becomes evident that if any three or four members are known, then the fourth can be determined. When the proportion given earlier, *a:b = c:d,* is written as a fraction, *a/b = c/d,* then $a \times d = c \times b$. If you substitute numbers for letters and for the proportion, then 1/2 = 2/4, $2 \times 2 = 4$, and $1 \times 4 = 4$. It is, therefore, obvious that the product of the means (2×2) equals the product of the extremes (1×4).

$$
\begin{array}{cc}
\text{extremes} & \text{extremes} \\
\downarrow \qquad \downarrow & \downarrow \qquad \downarrow \\
a:b = c:d & 1:2 = 2:4 \\
\uparrow \qquad \uparrow & \uparrow \qquad \uparrow \\
\text{means} & \text{means}
\end{array}
$$

Example 28

What is *x* in the proportion 2:3 = *x*:12?

Solution   Rewriting in fractional form,

$$\frac{2}{3} = \frac{x}{12}$$

Since $12 = 4 \times 3$, x must be 2×4, or 8, since this gives equal fractions. The proportion when completed is:

$$\frac{2}{3} = \frac{8}{12}$$

Another method of making use of the principle that the product of the means equals the product of the extremes is shown below:

$$2{:}3 = x{:}12$$

The product of the means, $3x$, equals the product of the extremes, 2×12 or 24. Thus $3x = 24$, and $1x$ must equal 8, the same answer obtained earlier.

Example 29

What is x in the proportion $5{:}x = 2{,}000{:}10{,}000$

Solution   Rewriting in fractional form.

$$\frac{5}{x} = \frac{2{,}000}{10{,}000}$$

And solve for the unknown value:

$$5 = \frac{(2{,}000)(x)}{10{,}000}$$

$$(5)(10{,}000) = (2{,}000)(x)$$

$$\frac{(5)(10{,}000)}{2{,}000} = x$$

$$25 = x$$

Example 30

What is a in the proportion $3{:}a = 6{:}20$?

Solution

$$\frac{3}{a} = \frac{6}{20}$$

$6 = 2 \times 3$; therefore, 20 must be twice the value of a, and a must then be 20/2, or 10. The complete proportion: $3/10 = 6/20$.

The alternative solution:

$$6a = 3 \times 20 \quad 6a = 60 \quad a = 10$$

The complete proportion:

$$3{:}10 = 6{:}20$$

Example 31

If a pump will fill a tank in 15 hours at 5 gpm (gallons per minute), how long will it take a 15-gpm pump to fill the same tank?

First, analyze the problem. Here the unknown is some number of hours. But should the answer be larger or smaller than 15 hours? If a 5-gpm pump can fill the tank in 15 hours, a larger pump (15-gpm) should be able to complete the filling in less than 15 hours. Therefore, the answer should be less than 15 hours. Now set up the proportion:

$$\frac{x \text{ hours}}{15 \text{ hours}} = \frac{5 \text{ gpm}}{15 \text{ gpm}}$$

$$x = \frac{(5)(15)}{(15)}$$

$$x = 5 \text{ hours}$$

It doesn't take long before you will gain an understanding of proportion problems that will allow you to skip some of the various steps to solving these problems (practice makes perfect and repetition aids easy recognition). In the following examples a short cut method is shown that will allow an experienced operator to solve problems quite easily.

Example 32

To make a certain chemical solution, 63.7 mg of a chemical must be added to 150 L of water. How much of the chemical should be added to 25 L to make up the same strength solution?

To solve this problem, you must first decide what is unknown, and whether you expect the unknown value to be larger or smaller than the known value of the same unit. The amount of chemical to be added to 25 L is the unknown, and you would expect this to be smaller than the 63.7 mg needed for 150 L.

First take the two known quantities of the same unit (25 L and 150 L) and make a fraction to multiply the third known quantity (63.7 mg) by. Notice that there are two possible fractions you can make with 25 and 150:

$$\frac{25}{150} \quad \text{or} \quad \frac{150}{25}$$

Second you want to choose the fraction that will make the unknown number of milligrams less than the known (63.7 mg). Multiplying 63.7 by the fraction 25/150 would result in a number *smaller* than 63.7. Multiplying 63.7 by 150/25, however, would result in a number larger than 63.7.

You wish to obtain a number smaller than 63.7, so multiply by 25/150; then complete the calculation to solve the problem:

$$\frac{(25)}{150}(63.7) = y$$

$$\frac{(25)(63.7)}{150} = y$$

$$10.62 \text{ mg} = y$$

From the above operation it should be obvious that the key to this method is arranging the two known values of like units into a fraction that, when multiplied by the third known value, will render a result that is smaller or larger as required.

Example 33

If a machine metal is composed of 25 parts copper and 15 parts tin, what is the weight of each in a machine weighing 2,200 lb?

Solution   The total weight of the machine, in parts, is equal to $15 + 25$, or 40 parts. The ratio of the number of parts of each metal to the total number of parts of metal equals the ratio of the weight of each metal to the total weight of metal. To calculate the pounds of copper, set up the following ratio:

$$\frac{25}{40} = \frac{c}{2,200} \quad \text{or} \quad 25:40 = c:2,200$$

Then

$$25 \times 2,200 = c \times 40$$

$$c = \frac{25 \times 2,200}{40}$$

$$c = 1,375 \text{ lb}$$

For tin,

$$\frac{15}{40} = \frac{t}{2,200} \quad \text{or} \quad 15:40 = t:2,200$$

$$15 \times 2,200 = t \times 40$$

$$t = \frac{15 \times 2,200}{40}$$

$$t = 825 \text{ lb}$$

3.9 FINDING AVERAGES

Finding *averages* or an arithmetic mean of a series of numbers is accomplished by adding the numbers and dividing by the number of numbers in the group. This is an activity that is required on several reports that are kept and maintained by water/wastewater operators (e.g, the monthly report on chlorine, sulfur dioxide, and turbidity readings). Typically, the average for the month is computed on all chemicals added, and on most test results.

Averaging plays another important role in water/wastewater treatment in that process performance can be assessed if enough data is collected and properly evaluated. Unfortunately, most data collected varies from time to time; thus, it is difficult to determine *trends* in process performance.

All is not lost however. This is the case because if you group the data (information) and then compute an average, a trend or trends may be determined. It is important to point out that an average is just that; that is, it is a reflection of the general nature of a certain group and does not necessarily reflect any one component of that group.

Let's look at a few examples.

Example 34

Find the average of the following series of numbers: 8, 12, 21, 12, 6, 9, 5, and 4.

Adding the numbers together, we get 77.

There are 8 numbers in this set.

Divide 77 by 8.

$$\frac{77}{8} = 9.6 \text{ is the average of the set}$$

Example 35

Let's try a series of daily turbidities. Obtain the average for them.

$$
\begin{array}{c}
0.9 \\
0.7 \\
0.1 \\
0.3 \\
0.4 \\
0.3 \\
0.4 \\
0.7
\end{array}
$$

The total is 3.8. There are 8 numbers in the set. Therefore:

$$\frac{3.8}{8} = 0.5\text{—rounding off}$$

3.10 PERCENT

Simply put, *percent* means "parts of 100 parts" and/or "by the hundred." The symbol for percent is %. Percent is used to describe portions of the whole. Thus 12% means 12 percent or 12/100 or 0.12. As another example, consider a tank that is 3/4 full; we say that it contains 75% of the original solution. Percent is also commonly used to describe the portion of a budget spent on a project completed. "There is only 10% of the budgeted amount remaining." "The treatment plant retro-fit is 30% complete."

Except when it is used in calculation, percentage is expressed as a whole number with a % sign after it. In a calculation percent is expressed as a decimal. The decimal is obtained by dividing the percent by 100. For example, 15% is expressed as the decimal 0.15, since 15% is equal to 15/100. This decimal is obtained by dividing 15 by 100.

To determine what percentage a part is of the whole divide the part by the whole: There are 120 sample bottles to clean; Robert has finished 30 of them. What percentage of the bottles have been cleaned?

(1) Step 1: $30 \div 120 = 0.25$
(2) Step 2: The 0.25 is converted to percent by multiplying the answer by 100.
(3) Step 3: $0.25 \times 100 = 25\%$. Thus, 25% of the 120 sample bottles have been cleaned.

To determine the whole when the part and its percentage is given, divide the part by the percentage. Example: How much 55% calcium hypochlorite is required to obtain 10 pounds of pure chlorine? The part is 10 pounds, which is 55% of the whole.

(1) Step 1: Convert the percentage to a decimal by dividing by 100.
$55\% \div 100 = 0.55$
(2) Step 2: Divide the part by the decimal equivalent of the percentage. 10 lbs $\div 0.55 = 18.2$ (rounded)

To change the percent obtained above to the decimal equivalent divide the percent by 100.
Change 40% to a decimal.

(1) Step 1: $40\% \div 100 = 0.40$ (0.40 is the decimal equivalent of 40%)

To find the percentage of a number multiply the number by the decimal equivalent of the percentage given in the problem. For example, what is 36% of 320?

(1) Step 1: Change the 36% to a decimal equivalent.
            $36\% \div 100 = 0.36$
(2) Step 2: Multiply $320 \times 0.36 = 115.2$
(3) Step 3: 36% of 320 is 115.2
            115.2 is 36% of 320

To increase a value by a percent we add the decimal equivalent of the percent to "1" and multiply it times the number.

A filter bed will expand 30% during backwash. If the filter bed is 48 inches deep, how deep will it be during backwash?

(1) Step 1: Change the percent to a decimal.
            $36\% \div 100 = 0.36$
(2) Step 2: Add the whole number 1 to this value.
            $1 + 0.36 = 1.36$
(3) Step 3: Multiply times the value, 48 in. $\times 1.36 = 65.3$ inches (rounded)

In water/wastewater treatment the concentration of chemicals used, such as sodium hydroxide and ferric chloride, is commonly expressed as a percentage. For example, a sulfur dioxide solution was made to have a 6% concentration. It is often desirable to determine this concentration in mg/L. To accomplish this we consider 6% as six percent of a million. [*Note:* A million because a liter of water weighs 1,000,000 mg. 1 mg in 1 liter is 1 part in a million parts (ppm)]. To find the concentration in mg/L when it is expressed in percent, do the following:

(1) Step 1: Change the percent to a decimal.
            $6\% \div 100 = 0.06$
(2) Step 2: Multiply times a million.
            $0.06 \times 1,000,000 = 60,000$ mg/L

3.10.1 PERCENT: PRACTICAL APPLICATIONS

The following examples illustrate the type of percentage problems that you will be required to calculate in water/wastewater operations.

Example 36

20% of the chlorine in a 40 gallon vat has been used. How many gallons are remaining in the vat?

(1) Step 1: Find the percentage of the chlorine remaining.
            $100\% - 20\% = 80\%$
(2) Step 2: Change the percent to a decimal.
            $80\% \div 100 = 0.80$

(3) Step 3: Multiply the percent as a decimal times the tank volume. 0.80×40 gal $= 32$ gal

Example 37

There are 60 pounds of pure chlorine in a drum. If the chlorine is 73% calcium hypochlorite, how much does the drum weigh?

(1) Step 1: Change the percent to a decimal.
$73\% \div 100 = 0.73$
(2) Step 2: Divide the weight by the percentage

$$\frac{60 \text{ lb}}{0.73} = 82.2 \text{ lb}$$

Example 38

A 3% chlorine solution is what concentration in mg/L?

(1) Step 1: Change the percentage to a decimal.
$3\% \div 100 = 0.03$
(2) Step 2: Multiply times one million.
$0.03 \times 1,000,000 = 30,000$ mg/L

Example 39

A water plant produces 93,000 gallons per day. 8,350 gallons are used to backwash the filter. What percentage of water is used to backwash?

(1) Step 1: Divide the part by the whole.

$$\frac{8,350 \text{ gal}}{93,000 \text{ gal}} = 0.09$$

(2) Step 2: Change the value to a percentage.
$0.09 \times 100 = 9\%$

Note: Water and wastewater treatment plants consist of a series of tanks and channels. Proper operational control requires you to be able to perform several process control calculations, including finding circumference, areas, volumes, and perimeters of a tank or channel as part of the data necessary to determine the result. To aid in performing these calculations the following definitions are provided.

• *Area*   the surface of an object, measured in square units.

- *Base*   the term used to identify the bottom leg of a triangle, measured in linear units.
- *Circumference*   the distance around an object, measured in linear units. When determined for other than circles it may be called the *perimeter* of the figure.
- *Cubic units*   measurements used to express volume, cubic feet, cubic meters, etc.
- *Depth*   the vertical distance from the bottom of the tank to the top. Normally measured in terms of water depth and given in terms of side wall depth (SWD), measured in linear units.
- *Diameter*   the distance from one edge of a circle to the opposite edge passing through the center, measured in linear units.
- *Height*   the vertical distance from the base or bottom of a unit to the top or water surface.
- *Length*   the distance from one end of the tank to the other, measured in linear units.
- *Linear units*   measurements used to express distances: feet, inches, meters, yards, etc.
- *Pi,π*   a number used in the calculations involving circles, spheres, or cones. $\pi = 3.1416$.
- *Radius*   the distance from the center of a circle to the edge, measured in linear units.
- *Sphere*   a container shaped like a ball.
- *Square units*   measurements used to express area, square feet, square meters, acres, etc.
- *Volume*   the capacity of the unit, how much it will hold, measured in cubic units (cubic feet, cubic meters) or in liquid volume units (gallons, liters, million gallons).
- *Width*   the distance from one side of the tank to the other, measured in linear units.

3.11 PERIMETER AND CIRCUMFERENCE

   As stated previously, water/wastewater treatment plants consist of a series of standard structures. These structures include contact tanks, clarifiers, troughs, aeration basins, sedimentation tanks, digesters, oxidation towers, and others. On occasion it is necessary to determine the distance around these structures and/or grounds (fenceline). In order to measure the distance around property, buildings and basin-/tank-like structures it is necessary to determine either perimeter or circumference. The *perimeter* is how far it is around an object, like a piece of ground. *Circumference* is the distance around a circle or circular structure. Distance is linear measurement, which defines the distance (or length) along a line. Standard units of measurement like inches, feet, yards, and miles, and metric units like centimeters, meters, and kilometers are used.

The perimeter of a rectangle (a 4-sided figure with 4 right angles) is obtained by adding the lengths of the four sides.

$$\text{Perimeter} = L_1 + L_2 + L_3 + L_4$$

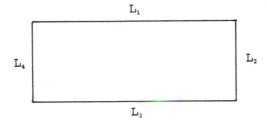

Find the perimeter of the following rectangle:

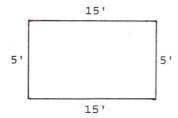

$$P = 15' + 5' + 15' + 5' = 40'$$

Example 40

You wish to put a fence around the chemical storage area at your plant. It has the dimensions shown below. Determine the perimeter of the area so you can order the correct length of fencing materials.

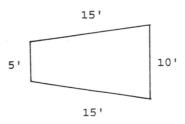

To calculate the perimeter, or distance around the area, add the lengths of all four sides:

$$P = 15' + 10' + 15' + 5'$$

$$P = 45'$$

The circumference of a circle is found by multiplying pi (π) times the diameter. (Diameter is a straight line passing through the center of a circle—the distance across the circle.)

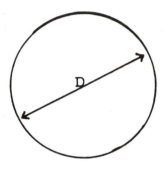

$$C = \pi D$$

where

C = circumference
π = Greek letter pi
π = 3.1416
D = diameter

Use this type of calculation if, for example, you must determine the circumference of a circular tank.

Find the circumference of a circle that has a diameter of 15 feet. (π = 3.14)

$$C = \pi \times 15' = 47.1 \text{ feet}$$

Example 41

A circular clarifier has a diameter of 18 m. What is the circumference of this tank?

$$C = \text{pi}\,(D)$$
$$C = (3.14)(\text{diameter})$$
$$= (3.14)(18\text{ m})$$
$$= 56.5\text{ m}$$

Example 42

A circular settling tank is 90 feet in diameter and has a weir around the outer edge of the tank. What is the circumference of the weir in feet?

$$C = \text{pi}\,(D)$$
$$C = 3.14 \times 90\text{ feet}$$
$$= 282.6\text{ feet}$$

3.12 AREA

For area measurements in water/wastewater plant calculations, three basic shapes are particularly important, namely circles, rectangles, and triangles.

Area is the amount of surface an object contains or the amount of material it takes to cover the surface. The area on top of an aeration basin is called the surface area. The area of the end of a collection of interceptor line (pipe) is called the *cross-sectional area* (the area at right angles to the length of a pipe or basin). Area is usually expressed in square units such as square inches (in.^{2}) or square feet (ft^2). Land may also be expressed in terms of square miles (sections) or acres (43,560 ft^2) or in the metric system as *hectares*.

The area of a rectangle is found by multiplying the length (L) times width (W).

$$\text{Area} = L \times W$$

L

W

Find the area of the following rectangle:

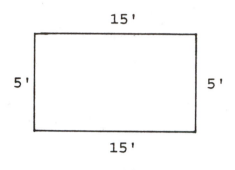

$$A = L \times W$$
$$= 15' \times 5'$$
$$= 75 \text{ ft}^2$$

The surface area of a circle is determined by multiplying pi times the radius squared.

Note: Radius, designated r, is defined as a line from the center of a circle or sphere to the circumference of the circle or surface of the sphere.

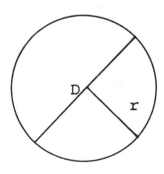

Area of circle $= \pi r^2$

where:

A = area
π = Greek letter pi ($\pi = 3.14$)
r = radius of a circle—radius is one-half the diameter

Example 43

What is the area of the circle shown below?

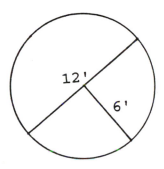

$$\text{Area of circle} = \pi r^2$$
$$= \pi 6^2$$
$$= 3.14 \times 36$$
$$= 113 \text{ ft}^2$$

Example 44

What is the area of the rectangle shown below?

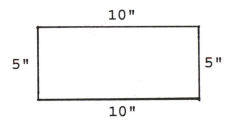

$$\text{Area of rectangle} = (\text{length})(\text{width})$$
$$= (10 \text{ in.})(5 \text{ in.})$$
$$= 50 \text{ in. surface area}$$

Note: It should be pointed out that even though area measurements are expressed in square units, this does *not* mean that the surface must be square in order to measure it. The point is that the surface of virtually any shape can be measured.

3.13 VOLUME

The amount of space occupied by or contained in an object, *volume*, is expressed in cubic units, such as cubic inches (in.^{3}), cubic feet (ft^3), acre feet (1 acre foot = 43,560 ft^3), etc.

The volume of a rectangular object is obtained by multiplying the length times the width times the depth or height.

$$V = L \times W \times H$$

where

L = length
W = width
D or H = depth or height

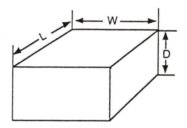

Example 45

Find the volume in cubic feet of an aeration basin with the following dimensions:

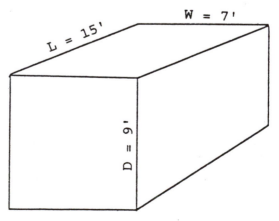

$$V = L \times W \times D$$
$$= 15' \times 7' \times 9'$$
$$= 945 \text{ ft}^3$$

For water/wastewater treatment plant calculations, representative surface areas are most often rectangles, triangles, circles, or a combination of these. Practical volume formulas used in treatment plant calculations are given in Table 3.2.

Example 46

Find the volume of a 4-inch pipe that is 300 feet long.
(1) Step 1: Change the diameter of the pipe from inches to feet by dividing by 12.

$$D = 4 \div 12 = 0.33 \text{ ft}$$

(2) Step 2: Find the radius by dividing the diameter by 2.

$$r = 0.33 \text{ ft} \div 2 = 0.165$$

(3) Step 3: Find the volume

$$V = L \times \pi r^2$$
$$V = 300 \text{ ft} \times \pi \times (.0225) \text{ ft}^2$$
$$V = 21.2 \text{ ft}^2$$

Example 47

Find the volume of a chemical vat that is 30 inches in diameter and 40 inches tall.
(1) Step 1: Find the radius of the vat. The radius is one-half the diameter.
30 inches $\div$ 2 = 15 inches

TABLE 3.2. Volume Formulas.

Sphere volume	=	$(\pi/6)(\text{diameter})^3$
Cone volume	=	1/3 (volume of a cylinder)
Rectangular tank volume	=	(area of rectangle)(D or H)
	=	(LW)(D or H)
Cylinder volume	=	(area of cylinder)(D or H)
	=	πr^2 (D or H)
Triangle volume	=	(area of triangle)(D or H)
	=	($bh/2$)(D or H)

Find the volume

$$V = H \times \pi r^2$$
$$V = 40 \text{ in.} \times \pi (15)^2$$
$$V = 40 \text{ in.} \times \pi (225 \text{ in.}^2)$$
$$V = 28,260 \text{ in.}^3$$

3.14 CONVERSION FACTORS

Conversion factors are used to change measurements or calculated values from one unit of measure to another. In making the conversion from one unit to another, you must know two things:

(1) The exact number that relates the two units
(2) Whether to multiply or divide by that number

For example, in converting from inches to feet, you must know that there are 12 in. in 1 ft, and you must know whether to multiply or divide the number of inches by 0.08333 (i.e., 1 in. = 0.08 ft).

When making conversions, there is often confusion about whether to multiply or divide; on the other hand, the number that relates the two units is usually known and thus is not a problem. In order to gain understanding of the proper methodology—"mechanics"—to use for various operations requires practice.

Along with using the proper "mechanics" and much practice in making conversions, probably the easiest and fastest method of converting units is to use a conversion table.

Note: Beginning with a discussion on how to make temperature conversions, the other sections to follow concentrate on the types of conversions that water/wastewater specialists deal with on a routine basis.

3.14.1 TEMPERATURE CONVERSIONS

In line with earlier statements that water is special, strange, and different, it is interesting to note that when water is in the presence of heat, it behaves (you guessed it) in a strange manner.

Molecularly speaking, water should freeze at −100°C and boil at −80°. Lucky for us, it doesn't. We are lucky because if water strictly adhered to these limits, our blood would boil in our bodies; furthermore, earth would be shrouded in superheated steam.

Have you ever looked at a pond in winter? If you have, you may have noticed that the pond contains water as solid, liquid, and gas, all at the same time. This phenomena occurs because water moves between 0° and +100°, passing through all three of its phases within the narrow, natural range of temperatures found on earth.

When working with substances that are not special, strange, and different (which takes in a large group of the other elements/compounds), you may have noted that the amount of heat needed to increase their temperature by one degree is the same with each degree involved. This is not the case with water. Water is most easily heated between 35 and 40°C (95 and 104°F). Coincidentally, this is also the narrow range of usual body temperature of most animals (especially the active ones), including man. The importance of this factor is difficult to pin down exactly, precisely; however, this coincidence could be important to life as we know it and to the environment that supports life. Simply stated, this coincidence may be vital—vital to the maintenance of all life on earth.

Getting back to conversion factors, a conversion is a number that is used to multiply or divide into another number in order to change the units of the number. In many instances, the conversion factor cannot be derived. It must be known. Therefore, tables are used to find the common conversions.

Many experienced operators memorize some standard conversions. This occurs as a result of on-the-job performance, practice, and repetition—not normally as a result of attempting to memorize them. Thus, there is a definite need for conversion tables. Now let's look at temperature conversions.

Most water/wastewater specialists are familiar with the formulas used for Fahrenheit and Celsius temperature conversions:

$$°C = 5/9(°F - 32°)$$

$$°F = 9/5(°C) + 32°$$

The difficulty arises when one is required to recall these formulas from memory. Probably the easiest way to recall these important formulas is to remember three basic steps for both Fahrenheit and Celsius conversions:

(1) Add 40°
(2) Multiply by the appropriate fraction (5/9 or 9/5)
(3) Subtract 40°

Obviously, the only variable in this method is the choice of 5/9 or 9/5 in the multiplication step. To make the proper choice, you must be familiar

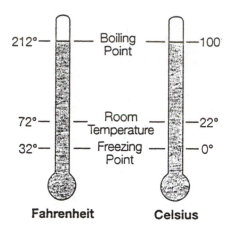

Figure 3.1 Fahrenheit and Celsius temperature scales.

with the two scales. As shown in Figure 3.1, on the Fahrenheit scale the freezing point of water is 32°, whereas it is 0° on the Celsius scale. The boiling point of water is 212° on the Fahrenheit scale and 100° on the Celsius scale.

What does all this mean?

Well, it is important to note, for example, that at the same temperature, higher numbers are associated with the Fahrenheit scale and lower numbers with the Celsius scale. This is an important relationship that helps you decide whether to multiply by 5/9 or 9/5. Let's look at a few conversion problems to see how the three-step process works.

Example 48

Suppose that you wish to convert 200°F to Celsius. Using the three-step process we proceed as follows:

(1) Step 1: add 40°

$$
\begin{array}{r}
200° \\
+40° \\
\hline
240°
\end{array}
$$

(2) Step 2: 240° must be multiplied by either 5/9 or 9/5. Since the conversion is to the *Celsius* scale, you will be moving to a number *smaller* than 240. Through reason and observation it is obvious that if 240 is multiplied by 9/5, it would almost be the same as multiplying by 2, which would double 240 rather than make it smaller. On the other hand, if you multiply by 5/9 it is about the same as multiplying by 1/2,

which would cut 240 in half. Since in this problem you wish to move to a smaller number, you should multiply by 5/9:

$$(5/9)(240°) = 1,200/9$$
$$= 133.3°C$$

(3) Step 3: Now subtract 40°:

$$\begin{array}{r} 133.3°C \\ -\ 40.0°C \\ \hline 83.3°C \end{array}$$

Therefore, 200°F = 83.3°C

Example 49

Convert 10°C to Fahrenheit
(1) Step 1: add 40°

$$\begin{array}{r} 10° \\ +40° \\ \hline 50° \end{array}$$

Since you are converting from Celsius to Fahrenheit, you are moving from a smaller to larger number, and 9/5 should be used in the multiplication:
(2) Step 2:

$$(9/5)(50°) = 450°/5$$
$$= 90°$$

(3) Step 3: Subtract 40°

$$\begin{array}{r} 90° \\ -40° \\ \hline 50° \end{array}$$

Thus, 10°C = 50°F.

Obviously, it is useful to know how to make these temperature conversion calculations. However, in practical plant operations you may wish to use a temperature conversion table.

3.14.2 OTHER IMPORTANT CONVERSIONS

As the old saying goes: To gain confidence in making standard conversions, the key to success is practice—practice—practice.

TABLE 3.3. Some Common Conversions.

Linear measurements	1 inch = 2.54 cm 1 foot = 30.5 cm 1 meter = 100 cm = 3.28 ft = 39.4 in. 1 acre = 43,560 ft² 1 yard = 3 feet
Volume	1 gal = 3.78 liters 1 ft³ = 1 liter 1 L = 1,000 mL 1 acre foot = 43,560 cubic feet 1 gal = 32 cups 1 pound = 16 oz dry wt
Weight	1 ft³ of water = 62.4 lb 1 gal = 8.34 lb 1 lb = 453.6 grams 1 kg = 1,000 g = 2.2 lb 1% = 10,000 mg/L
Pressure	1 ft of head = 0.433 psi 1 psi = 2.31 ft of heat
Flow	1 cfs = 448 gpm 1 gpm = 1,440 gpd

Table 3.3 shows common conversions used in water/wastewater practice. Table 3.4 shows conversions factors used to make the conversions. The information provided in this table will allow you to quickly make conversions.

The examples that follow illustrate many standard conversions that are made on a daily basis in water/wastewater operations.

Example 50

Cubic Feet to Gallons

$$\text{gallons} = \text{cubic feet, } (\text{ft}^3) \times 7.48 \text{ gal/ft}^3$$

How many gallons of biosolids can be pumped to a digester with 3,410 cubic feet of volume available?

$$\text{gallons} = 3,410 \text{ ft}^3 \times 7.48/\text{ft}^3 = 25,507 \text{ gallons}$$

Example 51

Gallons to Cubic Feet

$$\text{cubic feet (ft}^3) = \frac{\text{gallons}}{7.48 \text{ gal/ft}^3}$$

How many cubic feet of biosolids are removed when 14,500 gallons are withdrawn?

$$\text{cubic feet} = \frac{14,500 \text{ gallons}}{7.48 \text{ gal/ft}^3} = 1,939 \text{ ft}^3$$

Example 52

Gallons to Pounds

$$\text{pounds (lb)} = \text{gallons} \times 8.34 \text{ pounds/gallon}$$

If 1,250 gallons of solids are removed from the primary settling tank, how many pounds of solids are removed?

TABLE 3.4. Common Conversion Factors Used in Water/Wastewater Treatment.

To Change From	To	Multiply by
Feet (ft)	Inches (in.)	12
Yards (yd)	Feet (ft)	3
Yards (yd)	Inches (in.)	36
Miles (mi)	Yards (yd)	1,760
Miles (mi)	Feet (ft)	5,280
Cubic feet (ft³)	Gallons (gal)	7.48
Cubic feet (ft³)	Pounds-wastewater (lbs)	62.4
Cubic feet (ft³)	Pounds-air (lb)	0.075
Gallons (gal)	Pounds (lb)	8.34
Gallons (gal)	Liters (L)	3.785
Liters/second (L/s)	Gallons/minute (gpm)	15.85
Million gallons/day (MGD)	Gallons/day (gpd)	1,000,000
Million gallons/day (MGD)	Gallons/minute (gpm)	694
Million gallons/day (MGD)	Cubic feet/second (cfs)	1.55
Acres (Ac)	Square feet (ft²)	43,560
Acre feet (Ac-ft)	Gallons (gal)	325,829
Acre inches (Ac-in.)	Gallons (gal)	27,152
Liters (L)	Milliliters (ml)	1,000
Kilograms (kg)	Pounds (lb)	2.2
Meters (m)	Feet (ft)	3.3
Cubic meters (m³)	Gallons (gal)	269
Pounds (lb)	Grams (g)	454
Tons (t)	Pounds (lb)	2,000

To convert in the opposite direction divide by the factor.

pounds = 1,250 gallons × 8.34 pounds/gallon = 10,425

Example 53

Pounds to Gallons

$$\text{gallons (gal)} = \frac{\text{pounds (lb)}}{8.34 \text{ lb/gal}}$$

How many gallons of water are required to fill a tank which holds 6,500 lbs of water?

$$\text{gallons} = \frac{6,500 \text{ lb}}{8.34 \text{ pounds/gallon}} = 779 \text{ gallons}$$

Example 54

Milligrams/Liter to Pounds
Concentration in milligrams per liter or part per million (ppm) determined by laboratory testing must be converted to quantities in pounds, kilograms, pounds per day, or kilograms per day to be useful in controlling the plant.

pounds = concentration (mg/L) × volume (MG) × 8.34 lb/mg/L/MG

The solids concentration in the aeration tank is 2,500 mg/L. The aeration tank volume is 0.80 MG. How many pounds of solids are in the tank?

pounds = 2,500 mg/L × 0.80 MG × 8.34 lb/mg/L/MG = 16,680 lb

Example 55

Milligrams per Liter to Pounds per Day

pounds/day = concentration (mg/L) × flow, MGD × 8.34 lb/L/MG

How many pounds of solids are discharged per day when the plant effluent flow rate is 4.58 MGD and the effluent solids concentration is 25 mg/L?

pounds/day = 25 mg/L × 4.58 MGD × 8.34 lb/mg/L/MG = 954 lb/d

Example 56

Milligrams per Liter to Kilograms per Day

$$\text{kilograms/day} = \text{concentration, mg/L} \times \text{volume,}$$

$$\text{MG} \times 3.75 \text{ kg/mg/L/MG}$$

The effluent contains 25 mg/L of BOD_5. How many kilograms per day of BOD_5 are discharged when the effluent flow rate is 11.5 MGD?

$$\text{kilograms/day} = 25 \text{ mg/L} \times 11.5 \text{ MGD} \times 3.785 \text{ kg/mg/L/MG}$$

$$= 1,088 \text{ kg/d}$$

Example 57

Pounds to Milligrams per Liter

$$\text{concentration (mg/L)} = \frac{\text{quantity (lb)}}{\text{vol (MG)} \times 8.34 \text{ lb/mg/L/MG}}$$

The aeration tank contains 74,000 pounds of solids. The volume of the aeration tank is 3.15 MG. What is the concentration of solids in the aeration tank in mg/L?

$$\text{concentration (mg/L)} = \frac{74,000 \text{ lb}}{3.15 \text{ MG} \times 8.34 \text{ lb/MG/mg/L}} = 2,817 \text{ mg/L}$$

Example 58

Pounds per Day to Milligrams per Liter

$$\text{concentration (mg/L)} = \frac{\text{quantity (lb/day)}}{\text{flow (MGD)} \times 8.34 \text{ lb/mg/L/MG}}$$

The disinfection process uses 4,600 pounds per day of chlorine to disinfect a flow of 23 MGD. What is the concentration of chlorine applied to the effluent?

$$\text{concentration (mg/L)} = \frac{4,600 \text{ lb}}{23 \text{ MGD} \times 8.34 \text{ lb/MG/mg/L}} = 21.1 \text{ mg/L}$$

Example 59

Pounds to Flow in Million Gallons per Day

$$\text{flow (MGD)} = \frac{\text{quantity (lb/day)}}{\text{concentration (mg/L)} \times 8.34 \text{ mg/L/MG}}$$

You must remove 9,115 pounds of solids from the activated biosolids process per day. The waste activated biosolids concentration is 6,112 mg/L. How many million gallons per day of waste activated biosolids must be removed?

$$\text{flow (MGD)} = \frac{9,115 \text{ lb}}{6,112 \text{ mg/L} \times 8.34 \text{ lb/MG/mg/L}} = 0.178 \text{ MGD}$$

Example 60

Million Gallons per Day (MGD) to Gallons per Minute (gpm)

$$\text{flow (gpm)} = \frac{\text{flow (MGD)} \times 1,000,000 \text{ gallons/MG}}{1,440 \text{ minutes/day}}$$

The current flow is 3.85 MGD. What is the flow rate in gallons per minute?

$$\text{flow (gpm)} = \frac{3.85 \text{ MGD} \times 1,000,000 \text{ gallons/MG}}{1,440 \text{ minutes/day}} = 2,674 \text{ gpm}$$

Example 61

Million Gallons per Day (MGD) to Gallons per Day (gpd)

$$\text{flow (gpd)} = \text{flow (MGD)} \times 1,000,000 \text{ gallons/MG}$$

The influent meter reads 25.8 MGD. What is the current flow rate in gallons per day?

$$\text{flow (gpd)} = 25.8 \text{ MGD} \times 1,000,000 \text{ gal/MG} = 25,800,000 \text{ gpd}$$

Example 62

Million Gallons per Day (MGD) to Cubic Feet per Second (cfs)

$$\text{flow (cfs)} = \text{flow (MGD)} \times 1.55 \text{ cfs/MGD}$$

The flow rate entering the grit channel is 2.85 MGD. What is the flow rate in cubic feet per second?

$$\text{flow (cfs)} = 2.85 \text{ MGD} \times 1.55 \text{ cfs/MGD} = 4.42 \text{ cfs}$$

Example 63

Gallons per Minute (gpm) to Million Gallons per Day (MGD)

$$\text{flow (MGD)} = \frac{\text{flow (gpm)} \times 1,440 \text{ minutes/day}}{1,000,000 \text{ gallons/MG}}$$

The flow meter indicates that the current flow rate is 1,305 gpm. What is the flow rate in MGD?

$$\text{flow (MGD)} = \frac{1,305 \text{ gpm} \times 1,440 \text{ minutes per day}}{1,000,000 \text{ gal/MG}} = 1.88 \text{ MGD}$$

Example 64

Gallons per Day (gpd) to Million Gallons per Day (MGD)

$$\text{flow (MGD)} = \frac{\text{flow (gpd)}}{1,000,000 \text{ gallons/MG}}$$

The totalizing flow meter indicates that 33,550,500 gallons of wastewater have entered the plant in the past 24 hours. What is the flow rate in MGD?

$$\text{flow (MGD)} = \frac{33,550,500 \text{ gallons/day}}{1,000,000 \text{ gal/MG}} = 33.55 \text{ MGD}$$

Example 65

Flow in Cubic Feet per Second (cfs) to Million Gallons per Day (MGD)

$$\text{flow (MGD)} = \frac{\text{flow (cfs)}}{1.55 \text{ cfs/MG}}$$

The flow in a channel is determined to be 2.78 cubic feet per second (cfs). What is the flow rate in million gallons per day (MGD)?

$$\text{flow (MGD)} = \frac{2.78 \ (\text{cfs})}{1.55 \ \text{cfs/MG}} = 1.79 \ \text{MGD}$$

3.15 PRESSURE AND HEAD CALCULATIONS

Pressure is the force exerted on a unit area. Pressure = Weight (W) × Height (H). In water/wastewater, it is usually measured in psi (pounds per square inch). One foot of water exerts a pressure of 0.433 pounds per square inch. The pressure on the bottom of a container is not related to the volume of the container, nor the size of the bottom. The pressure is dependent on the height of the fluid in the container.

The height of the fluid in a container is referred to as *head*. More specifically, head is the measure of the pressure of water expressed as height of water in feet 1 psi = 2.31 feet of head.

Water weighs 62.4 pounds per cubic foot. The surface of any one side of a cube contains 144 square inches (12″ × 12″ = 144 in.²). Thus, we can say that the cube, if filled with water, contains 144 columns of water one foot tall and one inch square. The weight of each of these columns can be determined by dividing the weight of the water in the cube by the number of square inches.

$$\text{weight} = \frac{62.4 \ \text{lb}}{144 \ \text{in.}^2} = 0.433 \ \text{lb/in.}^2 \quad \text{or } 0.433 \ \text{psi}$$

0.433 pounds per square inch is the weight of one column of water one foot tall; thus, we can say that 0.433 pounds per square inch per foot of head or 0.433 psi/ft. In other words, 1 foot of head = 0.433 psi.

Now we can determine the relationship between pressure and head. That is, 1 foot of head represents a certain amount of head. This is determined by dividing 1 by 0.433 psi.

$$\text{feet of head} = \frac{1 \ \text{ft}}{0.433 \ \text{psi}} = 2.31 \ \text{ft/psi}$$

What we are really saying here is that if we were to read a pressure gauge and determine a reading of 15 psi, we know that the height of the water necessary to represent this pressure would be 15 psi × 2.31 ft/psi = 34.7 feet.

The point is that both conversions can be simply stated as

$$1 \text{ ft} = 0.433 \text{ psi}$$

and

$$1 \text{ psi} = 2.31 \text{ feet}$$

Note: When attempting to commit to memory mathematical conversions that are important (used often), it is sometimes confusing to try to memorize these two conversions; they mean the same thing. Therefore, it probably makes more sense to memorize one and stay with it. The most accurate conversion is the 1 ft = 0.433 psi. This is the conversion that is used in the examples shown below.

Example 66

Convert 50 psi to feet of head.

Solution

$$50 \text{ (psi/1)} \times \text{(ft/0.433)} =$$

$$\text{feet of head} = 115.5 \text{ feet}$$

Example 67

Convert 50 feet to psi.

Solution

$$50 \text{ (ft/1)} \times 0.433 \text{ psi/1 ft} = 21.7 \text{ psi}$$

Example 68

It is 115 feet in elevation between the top of the reservoir and the watering point. What will the static pressure be at the watering point?

Solution

$$115 \text{ (ft/1)} \times (0.433 \text{ psi/1 ft}) = 49.8 \text{ psi}$$

Example 69

A tank is 25 feet deep. What will be the pressure at the bottom of the tank?

Solution

$$25 \text{ (ft/1)} \times (0.433 \text{ psi/1 ft}) = 10.8 \text{ psi}$$

3.16 CALCULATIONS FOR FLOW

Determining the volume of water passing by a point over a period of time (*flow rate*) is an important calculation that is commonly used in water/wastewater treatment plant operations. Following are some of the units most frequently used for expressing flow rate.

- gallons per minute (gpm) or liters per minute (L/min)
- gallons per hour (gph) or liter per hour (L/h)
- gallons per day (gpd) or liters per day (L/d)
- million gallons per day (mgd) or megaliters per day (ML/d)
- cubic feet per second (ft³/s) or cubic meters per second (m³/s)

3.16.1 INSTANTANEOUS FLOW RATE

The rate at which flow (water/wastewater) is passing by a point at any instant is called the *instantaneous flow rate*. It should be pointed out that the instantaneous flow rate through a water/wastewater treatment plant varies constantly.

3.16.2 AVERAGE FLOW RATE

The average flow rate is found by dividing the total volume of water passing a point by the length of time. In water works, for example, the average use for each day is reported in gallons per minute (liters per minute). This is easily obtained by dividing the total volume of water pumped for the day by 1,440, the number of minutes in a day.

3.16.3 ANNUAL AVERAGE DAILY FLOW

When questions arise about whether or not a water system needs to be expanded, it is helpful to compute data that will support the argument for or against expansion. This can be accomplished by determining the *annual average daily flow*. Annual average daily flow is the average of average daily flows for a 12-month period. Usually measured in gpm, gph, gpd, or mgd, the annual average daily flow can be calculated by averaging the average daily flows for all the days of the year, or by dividing total flow volume for the year by 365.

3.16.4 PEAK-HOUR DEMAND

When attempting to determine the required capacity of distribution system piping, which ranges from 2.0 to 7.0 times the average hourly demand for the year, calculate the *peak-hour demand*. The peak-hour demand (greatest volume per hour flowing through the plant for any hour in the year) is determined from chart recordings showing the continuous changes in flow rates. Peak-hour demand is measured in gpm, gph, gpd, or mgd.

3.16.5 PEAK-DAY DEMAND

Peak-day demand is used to determine system operation during heaviest load period, which averages from 1.5 to 3.5 times the average daily flow rate. Peak-day demand is the greatest volume per day flowing through the plant for any day of the year. It is determined by looking at records of daily flow rates for the year and finding the peak day.

3.16.6 MINIMUM-DAY DEMAND

Minimum-day demand, which usually averages from 0.5 to 0.8 times the average flow rate, is used to determine possible plant shutdown periods. Simply put, minimum-day demand is the least volume per day flowing through the plant for any day of the year. In order to determine minimum-day demand look at records of daily flow rates for the year to find the minimum day.

3.16.7 PEAK-MONTH DEMAND

Peak-month demand determines possible plant shortage needs, which range from 1.1 to 1.5 times the average monthly flow volume for the year. Peak-month demand is the greatest volume passing through the plant during a calendar month. It is calculated by looking at records of daily flow volumes (totalizer records) to find peak month for the year. Peak-month demand is measured in gallons, millions of gallons, and/or billions of gallons.

3.16.8 MINIMUM-MONTH DEMAND

Minimum-month demand determines the best time to perform preventive maintenance and repairs to equipment because it measures the least volume passing through the plant during a calendar month. To calculate minimum-month demand look at records of monthly flow volumes to find minimum month for the year.

Examples of various flow problems are presented in the following section.

Example 70

Find the flow in MGD when the flow is 41,000 gpd.

Solution

$$\frac{41,000}{1,000,000} = 0.041 \text{ MGD}$$

Example 71

Find the flow in gpm when the total flow for the day is 85,000 gpd.

Solution

$$\frac{85,000 \text{ gpd}}{1,440 \text{ min/day}} = 59 \text{ gpm}$$

Example 72

Find the flow in gpm when the flow is 1.6 cfs.

Solution

$$1.6 \, (\text{cfs}/1) \times 448 \text{ gpm}/1 \text{ cfs} = 717 \text{ gpm}$$

Example 73

Find the flow in a 3-inch pipe when the velocity is 1.2 feet per second.

Solution

(1) Step 1: The diameter of the pipe is 3 inches. Therefore the radius is 1.5 inches. Convert the 1.5 inches to feet

$$3/12 = 0.25 \text{ ft}$$

(2) Step 2: $A = \pi\, r^2$
$\qquad A = \pi(0.25 \text{ ft})^2$
$\qquad A = \pi \times 0.0625 \text{ ft}^2$
$\qquad A = 0.20 \text{ ft}^2$

(3) Step 3: $Q = VA$

where:

Q = flow rate
A = area
V = velocity

$$Q = 1.2 \text{ ft /sec} \times 0.20 \text{ ft}^2$$

$$= 0.24 \text{ cfs}$$

3.17 DETENTION TIME

The term, *detention time* or *hydraulic detention time*, refers to the average length of time a drop of water, wastewater, or suspended particles remains in a tank or channel. It is calculated by dividing the water/wastewater in the tank by the flow rate through the tank. The units of flow rate used in the calculation are dependent on whether the detention time is to be calculated in minutes, hours, or days. Detention time is used in conjunction with various treatment processes, including sedimentation and coagulation-flocculation.

Generally, in practice, detention time is associated with the amount of time required for a tank to empty. The range of detention time varies with the process. For example, in a tank used for sedimentation, detention time is commonly measured in minutes.

The calculation method used to determine detention time is illustrated in the following examples.

Example 74

For Detention Time in Days

$$\text{DT (days)} = \frac{\text{tank volume (ft}^3 \times 7.48 \text{ gal /ft}^3}{\text{flow (gallons /day)}}$$

An anaerobic digester has a volume of 3,000,000 gallons. What is the detention time in days when the influent flow rate is 0.75 MGD?

Solution

$$DT\ (days) = \frac{3,000,000\ gallons}{0.75\ MGD \times 1,000,000\ gallons/MG}$$

$$DT\ (days) = 4\ days$$

Example 75

Detention Time in Hours

$$DT\ (hours) = \frac{tank\ volume\ (ft^3 \times 7.48\ gal/ft^3 \times 24\ hours/day}{flow\ (gallons/day)}$$

A settling tank has a volume of 45,000 ft³. What is the detention time in hours when the flow is 4.0 MGD?

$$DT\ (hours) = \frac{45,000\ ft^3 \times 7.48\ gal/ft^3 \times 24\ hr/day}{4.0\ MGD \times 1,000,000\ gal/MG}$$

$$DT\ (hours) = 2.0\ hours$$

Example 76

Detention Time in Minutes

$$DT\ (minutes) = \frac{tank\ volume\ (ft^3 \times 7.48\ gal/ft^3 \times 1,440\ minutes/day}{flow\ (gallons/day)}$$

A grit channel has a volume of 1,250 ft³. What is the detention time in minutes, when the flow rate is 4.0 MGD?

$$detention\ time\ in\ minutes = \frac{1,250\ ft^3 \times 7.48\ gal/ft^3 \times 1,440\ min/day}{4,000,000\ gallons/day}$$

$$= 3.37\ minutes$$

CHAPTER 4

Water Hydraulics

Anyone who has tasted natural spring water knows that it is different from city water, which is used over and over again, passing from mouth to laboratory and back to mouth again, without ever being allowed to touch the earth. We need to practice such economics these days, but in several thirsty countries, there are now experts in hydrodynamics who are trying to solve the problem by designing flowforms that copy the earth, producing rhythmic and spiral motions in moving water. And these pulsations do seem to vitalize and energize the liquid in some way, changing its experience, making it taste different and produce better crops. (Watson, 1988, p. 138)

4.1 HYDRAULICS

HYDRAULICS is the study of how liquids act as they move through a channel or a pipe. Hydraulics plays an important role in the design and operation of both the wastewater collection/water distribution systems and the treatment plants. The following section presents information about several basic concepts of hydraulics.

4.1.1 TERMINOLOGY

- *Friction head* — the energy needed to overcome friction in the piping system. It is expressed in terms of the added system head required.
- *Head* — the equivalent distance water must be lifted to move from the supply tank or inlet to the discharge. Head can be divided into three components: *static head, velocity head,* and *friction head.*
- *Pressure* — the force exerted per square unit of surface area. May be expressed as pounds per square inch.
- *Static head* — the actual vertical distance from the system inlet to the highest discharge point.

83

- *Total dynamic head*   the total of the static head, friction head, and velocity head.
- *Velocity*   the speed of a liquid moving through a pipe, channel, or tank. May be expressed in feet per second.
- *Velocity head*   the energy needed to keep the liquid moving at a given velocity. It is expressed in terms of the added system head required.

4.1.2 BASIC CONCEPTS

$$1 \text{ ft}^3 \text{ H}_2\text{O} = 62.4 \text{ lb}$$

The relationship shown above is important. The point is that cubic feet and pounds are both used to describe a volume of water. There is a defined relationship between these two methods of measurement. The specific weight of water is defined relative to a cubic foot. One cubic foot of water weighs 62.4 pounds. This relationship is true only at a temperature of 4°C and at a pressure of one atmosphere (called standard temperature and pressure—14.7 lbs per square inch at sea level). The weight varies so little, however, that for practical purposes we use this weight from a temperature 0°C to 100°C. The important point being made here is that

$$1 \text{ ft}^3 \text{ H}_2\text{O} = 62.4 \text{ lb}$$

A second relationship is also important:

$$1 \text{ gallon H}_2\text{O} = 8.34 \text{ pounds}$$

At standard temperature and pressure one cubic foot of water contains 7.48 gallons. With these two relationships we can determine the weight of one gallon of water. This is accomplished by

$$\text{wt. of gallon of water} = 62.4 \text{ lb}/7.48 \text{ gal} = 8.34 \text{ lb/gal}$$

Thus, the key point is

$$1 \text{ gallon H}_2\text{O} = 8.34 \text{ pounds}$$

Further, this information allows us to convert cubic feet to gallons by simply multiplying the number of cubic feet by 7.48 gal/ft^3.

Example 77

Find the number of gallons in a reservoir that has a volume of 707.9 ft³.

$$707.9 \text{ ft}^3 \times 7.48 \text{ gal/ft}^3 = 5,295 \text{ gallons}$$

4.1.3 PROPERTIES OF WATER: TEMPERATURE/SPECIFIC WEIGHT/DENSITY

Table 4.1 is provided to show the relationship between temperature, specific weight and density.

4.1.4 DENSITY AND SPECIFIC GRAVITY

When we say that steel is heavier than aluminum, we say that steel has greater density than aluminum. In practice, what we are really saying is that a given volume of steel is heavier than the same volume of aluminum.

What is density? *Density* is the *mass per unit volume* of a substance. Suppose you had a cube of butter and a large box of cereal, each having a

TABLE 4.1.

Temperature (°F)	Specific Weight (lb/ft³)	Density (slugs/ft³)
32	62.4	1.94
40	62.4	1.94
50	62.4	1.94
60	62.4	1.94
70	62.3	1.94
80	62.2	1.93
90	62.1	1.93
100	62.0	1.93
110	61.9	1.92
120	61.7	1.92
130	61.5	1.91
140	61.4	1.91
150	61.2	1.90
160	61.0	1.90
170	60.8	1.89
180	60.6	1.88
190	60.4	1.88
200	60.1	1.87
210	59.8	1.86

mass of 500 grams. The density of the cereal would be much less than the density of the butter because the cereal occupies a much larger volume than the butter occupies.

The density of an object can be calculated by using the formula:

$$\text{density} = \frac{\text{mass}}{\text{volume}}$$

In the water/wastewater fields, perhaps the most common measures of density are pounds per cubic foot (lb/ft³) and pounds per gallon (lb/gal). The density of a dry material, such as cereal, sand, lime, and soda ash, is usually expressed in pounds per cubic foot. The density of a liquid, such as liquid alum, liquid chlorine, or water, can be expressed either as pounds per cubic foot or as pounds per gallon. The density of a gas, such as chlorine gas, methane, carbon dioxide, or air, is usually expressed in pounds per cubic foot.

As shown in Table 4.1, the density of a substance like water changes slightly as the temperature of the substance changes. This happens because substances usually increase in volume (size) as they become warmer. Because of this expansion with warming, the same weight is spread over a larger volume, so the density is lower when a substance is warm than when it is cold.

What is specific gravity? *Specific gravity* is the weight of a substance compared to the weight of an equal volume of water. This relationship is easily seen when a cubic foot of water, which weighs 62.4 pounds, is compared to a cubic foot of aluminum, which weighs 178 pounds. Aluminum is 2.7 times as heavy as water.

It is not that difficult to find the specific gravity of a piece of metal. All you have to do is to weigh the metal in air, then weigh it under water. Its loss of weight is the weight of an equal volume of water. To find the specific gravity divide the weight of the metal by its loss of weight in water.

$$\text{specific gravity} = \frac{\text{weight of a substance}}{\text{weight of equal volume of water}}$$

Example 78

Suppose a piece of metal weighs 110 pounds in air and 72 pounds under water. What is the specific gravity?

(1) Step 1: 110 lb subtract 72 lb = 38 lb loss of weight in water

(2) Step 2:

$$\text{specific gravity} = \frac{110 \text{ lb}}{38 \text{ lb}} = 2.9$$

Note: In a calculation of specific gravity, it is *essential* that the densities be expressed in the same units.

The specific gravity of water is one, which is the standard, the reference for which all other substances are compared. That is, any object that has a specific gravity greater than one will sink in water. Considering the total weight and volume of a ship, its specific gravity is less than one; therefore, it can float.

The most common use of specific gravity in water/wastewater treatment operations is in gallons-to-pounds conversions. In many such cases, the liquids being handled have a specific gravity of 1.00 or very nearly 1.00 (between 0.98 and 1.08), so 1.00 may be used in the calculations without introducing much error. However, in calculations involving a liquid with a specific gravity less than 0.98 or greater than 1.02, the conversions from gallons to pounds must take specific gravity into account. The technique is illustrated in the following example.

Example 79

There are 1,305 gal of a certain liquid in a basin. If the specific gravity of the liquid is 0.91, how many pounds of liquid are in the tank?

Normally, for a conversion from gallons to pounds, we would use the factor 8.34 lb/gal (the density of water) if the substance's specific gravity were between 0.98 and 1.02. However, in this example the substance has a specific gravity outside this range, so the 8.34 factor must be adjusted. Multiply 8.34 lb/gal by the specific gravity to obtain the adjusted factor:

(1) Step 1: (8.34 lb/gal) (0.91) = 7.59 lb/gal
(2) Step 2: Then convert 1,305 gal to pounds using the corrected factor:

$$(1,305 \text{ gal}) (7.59 \text{ lb/gal}) = 9,905 \text{ lb}$$

4.1.5 FORCE AND PRESSURE

Force can be defined as the push or pull influence that causes motion. In the English system, force and weight are often used in the same way. The weight of a cubic foot of water is 62.4 pounds. The force exerted on the bottom of a one foot cube is 62.4 pounds. If we stack two cubes on top of one another, the force on the bottom will be 124.8 pounds.

Pressure is a force per unit of area. Pounds per square inch or pounds per square foot are common expressions of pressure. The pressure on the bottom of a cube is 62.4 pounds per square foot. It is normal to express pressure in pounds per square inch (psi). This is easily accomplished by determining the weight of one square inch of a cube one foot high. If we have a cube that is 12 inches on each side, the number of square inches on the bottom surface of the cube is $12 \times 12 = 144$ in². Now by dividing the weight by the number of square inches we can determine the weight on each square inch.

$$\text{psi} = \frac{62.4 \text{ lb/ft}}{144 \text{ in.}^2} = 0.433 \text{ psi/ft}$$

This is the weight of a column of water one inch square and one foot tall. If the column of water were two feet tall, the pressure would be 2 ft × 0.433 psi/ft = 0.866.

Key Point: 1 ft of water = 0.433 psi

With the above information we can convert feet of head (to be discussed in detail later) to psi by multiplying the feet of head times 0.433 psi/ft.

Example 80

A tank is mounted at a height of 50 feet. Find the pressure at the bottom of the tank.

Solution

50 ft × 0.433 psi/ft = 21.7 psi

If you wanted to make the conversion of psi to feet you would divide the psi by 0.433 psi/ft.

Example 81

Find the height of water in a tank if the pressure at the bottom of the tank is 15 psi.

Solution

$$\text{feet} = \frac{15 \text{ psi}}{0.433 \text{ psi/ft}} = 34.6 \text{ feet}$$

4.1.6 HEAD

Pressure is directly related to the height of a column of fluid. Put another way, *head* is the vertical distance the water/wastewater must be lifted from a supply source to discharge.

It is interesting to note that the pressure at the bottom of a container is only affected by the height of water in the container and not by the shape of the container. Figure 4.1 shows three containers of different shapes and sizes. The pressure at the bottom of each is the same.

The total head includes the vertical distance the liquid must be lifted (static head), the loss of friction (friction head), and the energy required to maintain the desired velocity (velocity head).

Key Point: total head, ft = static head + friction head + velocity head

The pressure exerted at the bottom of a tank is relative only to the head on the tank and not the volume of water in the tank.

4.1.6.1 Static Head

Static head is the actual *vertical* distance the liquid must be lifted.

Key Point: static head, ft = discharge elevation,

ft – supply elevation, ft

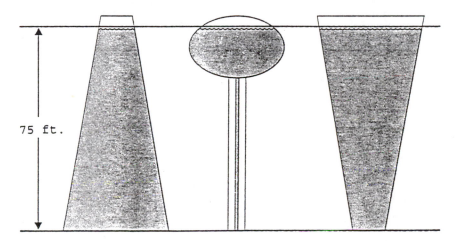

75 ft.

Figure 4.1 Three containers of different shapes and sizes. The pressure at the bottom of each is the same.

Example 82

The supply tank is located at elevation 112 feet. The discharge point is at elevation 215 feet. What is the static head in feet?

$$\text{static head} = 215 \text{ ft} - 112 \text{ ft} = 103 \text{ feet}$$

4.1.6.2 Friction Head

Friction head is the equivalent distance of the energy which must be supplied to overcome friction. Engineering texts include tables showing the equivalent vertical distances for various sizes and types of pipes, fittings, and valves. The total friction head is the sum of the equivalent vertical distances for each component.

Key Point: friction head, ft = energy losses due to friction

4.1.6.3 Velocity Head

Velocity head is the equivalent distance of the energy consumed in achieving and maintaining the desired velocity in the system.

Key Point: velocity head, ft = energy losses to maintain velocity

4.1.6.4 Total Dynamic Head (Total System Head)

Total Head = static head + friction head + velocity head
Static Head ft = discharge elevation – supply elevation
Friction Head, ft = energy losses due friction
Velocity Head, ft = energy losses to maintain velocity

4.1.6.5 Pressure/Head

The pressure exerted by water/wastewater is directly proportional to its depth or head in the pipe, tank, or channel. If the pressure is known, the equivalent head can be calculated.

$$\text{pressure, psi} = \frac{\text{head, ft}}{2.31 \text{ ft/psi}}$$

Example 83

The pressure gauge on the discharge line from the influent pump reads 76.5 psi. What is the equivalent head in feet?

Head, ft = 76.5 psi × 2.31 ft/psi = 176.7 ft

4.1.6.6 Head/Pressure

If the head is known, the equivalent pressure can be calculated by:

$$\text{pressure, psi} = \frac{\text{head, ft}}{2.31 \text{ ft/psi}}$$

Example 84

The tank is 15 feet deep. What is the pressure in psi at the bottom of the tank when it is filled with water?

$$\text{pressure, psi} = \frac{15 \text{ ft}}{2.31/\text{psi}} = 6.49 \text{ psi}$$

4.1.6.7 Flow, Area, and Velocity

The flow rate through an open channel is directly related to the liquid and the cross-sectional area of the liquid in the channel.

Key Point: flow (Q) = area (A) × velocity (V)

Example 85

A channel is 5 feet wide and the water depth is 2 feet. The velocity in the channel is 4 feet per second. What is the flow rate in cubic feet per second?

flow = 5 ft × 2 ft × 4 ft/second

= 40 cfs

Note: One cubic foot per second is equal to 448 gallons per minute.

4.1.7 MEASURING FLOW

It is important to determine, at any given time, the flow rate through the water/wastewater treatment plant. It is also important to be able to record these flow rates. Most devices used for measuring flow rate can be provided with equipment that will give a continuous recording of flow rates entering or leaving the treatment plant.

Flow rates in channels are measured by *weirs* and *Parshall flumes*. Weirs are of two common types, the *V-notch* and *rectangular* weirs. The flow rate is read using a graph or table pertaining to a weir. In using a graph or table to determine the flow rate, you must know two measurements: (1) the height (*H*) of the water above the weir crest; and (2) the angle of the weir (V-notch weir) or the length of the crest (rectangular weir).

The Parshall flume is also used to measure flows in open channels. A *nomograph* is used to determine flow rate in a Parshall flume. In order to use the nomograph you must know the width of the throat section of the flume (*W*), the upstream depth of the water (depth H_1 in the tube of one stilling well), and the depth of water at the front of the throat section (depth H_2 in the tube of the other stilling well).

4.1.8 PIEZOMETRIC SURFACE

The surface of water that is in contact with the atmosphere is known as *free water surface*. Many important hydraulic measurements are based on the difference in height between the free water surface and some point in the water system. The *piezometric surface* is used to locate this free water surface in a vessel, where it cannot be observed directly. The piezometric surface is an imaginary surface that coincides with the level of the water in an aquifer, or the level to which water in a system would rise in a *piezometer* (an instrument used to measure pressure). To understand how a piezometer actually measures pressure consider the following example.

If you connect a clear, see-through straw to the side of a clear plastic cup, the water will rise in the straw to indicate the level of the water in the cup. Such a see-through straw, the piezometer, allows you to see the level of the top of the water in the tube; this is called the piezometric surface.

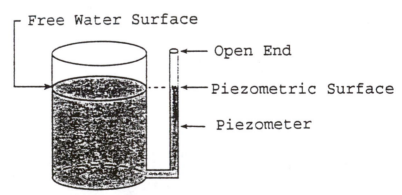

Figure 4.2 A container not under pressure where the piezometric surface is the same as the free water surface in the vessel.

Pressure Applied

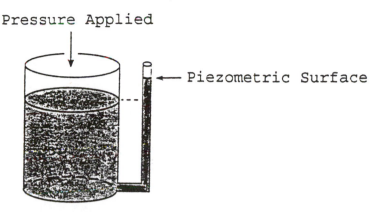 ←— Piezometric Surface

Figure 4.3 A container under pressure where the piezometric surface is above the level of the water in the tank.

In practice, a piezometer is connected to the side of a tank or pipeline. If the water-containing vessel is not under pressure (as is the case in Figure 4.2), the piezometric surface will be the same as the free water surface in the vessel, just as it would if a drinking straw (the piezometer) were left standing in a glass of water.

In a tank and pipeline system, when pressurized, as they often are, the pressure will cause the piezometric surface to rise above the level of the water in the tank. The greater the pressure, the higher the piezometric surface (Figure 4.3). An increased pressure in a water pipeline system is usually obtained by elevating the water tank.

To this point we have seen that water always rises to the water level of the main body of water. It should be pointed out, however, that this is only when the water is at rest (static or standing water). The situation is quite different when water is flowing. Consider, for example, an elevated storage tank feeding a distribution system pipeline. When the system is at rest, all valves closed, all the piezometric surfaces are the same height as the free water surface in storage. On the other hand, when the valves are opened, the water begins to flow and the piezometric surfaces changes. This is an important point because as water continues to flow down a pipeline, less and less pressure is exerted. This is the case because some pressure is lost (used up) keeping the water moving over the interior surface of the pipe (friction). The pressure that is lost is called *head loss*.

CHAPTER 5

Water Chemistry

Chemical testing can be divided into two types. The first type measures a bulk physical property of the sample, such as volume, temperature, melting point, or mass. These measurements are normally performed with an instrument, and one simply has to calibrate the instrument to perform the test. Most analyses, however, are of the second type, in which a chemical property of the sample is determined that generates information about how much of what is present. (Smith, 1995, p. 1)

5.1 INTRODUCTION

WHEN it comes to actually studying water at its most basic, elemental level, you first must recognize that no one has even seen a molecule of water. All that is available to us is equations and theoretical diagrams. When we look at the H_2O formula, we instantly think that water is simple. It is a mistake, of course, to think of water as being simple; it is not—it is very complex.

Although no one has seen a water molecule, we have determined through x-rays that atoms in water are elaborately meshed. Moreover, although it is true that we do not know as much as we need to know about water, we have determined many things about water. A large amount of our current knowledge comes from studies of water chemistry.

Water chemistry is important. This is especially the case for water/wastewater specialists. Water chemistry is important because several important factors about water that is to be treated and then distributed are determined through simple chemical analysis. Probably the most important determination that the water specialist makes about water is its hardness. The wastewater specialist, on the other hand, uses chemistry to determine other factors. For example, the wastewater operator may be interested in some of the same chemical results as water specialists but also needs to determine the levels of organics in the waste stream.

Why chemistry? "I'm not a chemist," you say.

Well, simply put, when you add chlorine to water to make it safe to drink or safe to deposit in a receiving body (usually a river or lake), you are a chemist. Chemistry is the study of substances and the changes they undergo. Water, wastewater, and/or other water specialists must possess a fundamental knowledge of chemistry. Therefore, this chapter covers the fundamentals of chemistry specific to water or wastewater practices.

Before beginning our discussion of water chemistry it is important for the reader to have some basic understanding of chemistry and concepts. Thus, the following chemistry terms, definitions, and concepts are presented for review and/or understanding.

5.1.1 CONCEPTS: "MISCIBLE," "SOLUBILITY," "IN SOLUTION," "DISSOLVED"

(1) Miscible: capable of being mixed in all proportions. Simply stated: when two or more substances that disperse themselves uniformly in all proportions when brought into contact are said to be completely soluble in one another, or completely miscible. The precise chemistry definition is: "homogenous molecular dispersion of two or more substances" (Jost, 1992, p. 101). Examples are:
   • All gases are completely miscible.
   • Water and alcohol are completely miscible.
   • Water and mercury (in its liquid form) are immiscible liquids.
(2) Between the two extremes of miscibility, there is a range of solubility; that is, various substances mix with one another up to a certain proportion. In many environmental situations, a rather small amount of contaminant is soluble in water in contrast to complete miscibility of water and alcohol. The amounts are measured in part per million.

5.1.2 CONCEPTS: "SUSPENSION," "SEDIMENT," "PARTICLES," "SOLIDS"

(1) Often water carries solids or particles in suspension. These dispersed particles are much larger than molecules and may be comprised of millions of molecules. The particles may be suspended in flowing conditions and initially under quiescent conditions; but eventually gravity causes settling of the particles. The resultant accumulation by settling is often called sediment or biosolids (sludge) or residual solids in wastewater treatment vessels. Between this extreme of ready falling out by gravity and permanent dispersal as a solution at the molecular level, there are intermediate types of dispersion or suspension. Particles can be so finely milled or of such small intrinsic size as to remain in suspension almost indefinitely and in some respects similarly to solutions.

5.1.3 EMULSION

(1) *Emulsions* represent a special case of a suspension. As you know, oil and water do not mix. Oil and other hydrocarbons derived from petroleum generally float on water with negligible solubility in water. In many instances oils may be dispersed as fine oil droplets (an emulsion) in water and not readily separated by floating because of size and/or the addition of dispersal promoting additives. Oil and, in particular, emulsions can prove detrimental to many treatment technologies and must be treated in the early steps of a multistep treatment process.

5.1.4 ION

(1) An *ion* is an electrically charged particle. For example, sodium chloride or table salt forms charged particles on dissolution in water; sodium is positively charged (a cation), and chloride is negatively charged (an anion). Many salts similarly form cations and anions on dissolution in water (Quagliano, 1964).

5.2 THE STRUCTURE OF MATTER

Matter is defined as anything that has weight (mass) and occupies space. When you think about, it does not take long to realize the obvious; that is, it is virtually impossible to name all those things that are classified as matter. There are too many kinds of matter. For example, a short list might include such items as water, air, kerosene, chlorine, rocks, tea, paper, and others.

It is safe to say (quite safe) that matter is just about everywhere. It's also safe to say (and quite obvious) that all matter is not the same. Matter can, however, be put into three large groups or *physical states of matter:* solids, liquids, and gases.

(1) *Solids* have definite shape with their particles closely packed together and sticking firmly to each other.
(2) *Liquids* change shape to fit their container but will maintain a constant volume. The components (particles) of the liquid move freely over one another but tend to stick together enough to maintain a constant volume.
(3) *Gases* have no fixed shape and their volume changes; they expand or compress to fill different sized containers. The components that make up a gas are free to move about; they do not stick together.

5.2.1 CHANGING MATTER

Changes in matter occur either physically or chemically. When matter changes size, shape, density, etc., as well as when gas changes its state, i.e., from gas to a liquid to a solid, this is a *physical change.* When a gas changes to form a new substance, e.g., iron rusting, this is a *chemical change.*

5.2.2 ATOMIC STRUCTURE OF WATER

Matter is composed of pure basic substances. Sometimes these basic substances are combined to form matter. This basic substance, an *element* or *elements,* is/are incapable of being broken down into simpler substances. There are more than 100 known elements. Ranging from simple to the very complex, elements exist in nature in both pure and combined forms. Some of the pure elements that are used in water/wastewater process operations are calcium, chlorine, oxygen, and carbon.

The smallest unit of an element is the *atom.* The most simple atom possible consists of a nucleus having a single proton with a single electron traveling around it. This is an atom of hydrogen, which has an atomic weight of one because of the single proton. The *atomic weight* of an element is equal to the total number of protons and neutrons in the nucleus of an atom of an element.

In order to gain an understanding of basic atomic structure and related chemical principles it is useful to compare the atom to our solar system. In our solar system the sun is the center of everything. The *nucleus* is the center in the atom. The sun has several planets orbiting around it. The atom has *electrons* orbiting about the nucleus. It is interesting to note that the astrophysicist, who would likely find this analogy simplistic, is concerned mostly with activity within the nucleus. This is not the case, however, with the chemist. The chemist deals principally with the activity of the planetary electrons; chemical reactions between atoms or molecules involve only electrons, with no changes in the nuclei.

The nucleus is made up of positive electrically charged *protons* and *neutrons,* which are neutral (no charge). The positive charge in the nucleus is balanced by the negatively charged electrons orbiting it. An electron has negligible mass (less than 0.02% of the mass of a proton), which makes it practical to consider the weight of the atom as the weight of the nucleus.

Atoms are identified by name, atomic number, and atomic weight. The *atomic number* or *proton number* is the number of protons in the nucleus of an atom. It is equal to the positive charge on the nucleus. In a neutral atom, it is also equal to the number of electrons surrounding the nucleus. As stated previously, the atomic weight of an atom depends on the number of protons and neutrons in the nucleus, the electrons having negligible mass. Atoms (elements) received their names and symbols in interesting ways. The discoverer of an atom usually proposes a name for it. Some atoms get their symbols from languages other than English. The following is a list of common elements with their common names and the names from which the symbol is derived.

- chlorine   Cl
- copper   Cu (*Cuprum*—Latin)
- hydrogen   H
- iron   Fe (*Ferrum*—Latin)
- nitrogen   N
- oxygen   O
- phosphorus P
- sodium   Na (*Natrium*—Latin)
- sulfur   S

As shown above, each atom is designated by a capital letter or a capital letter and a small letter. These are called chemical symbols. As is apparent from the above list, most of the time the symbol is easily recognized as an abbreviation of the atom name, such as O for oxygen.

5.2.3 ATOMS AND MOLECULES

Most of the elements are not found as single atoms all by themselves. They are more often found in combinations of atoms called *molecules*. Basically, a molecule is the least common denominator of making a substance what it is.

A system of formulas has been devised to show how atoms are combined into molecules.

When a chemist writes the symbol for an element, it stands for one atom of that element. A subscript following the symbol indicates the number of atoms in the molecule. O_2 is the chemical formula for an oxygen molecule. It shows that oxygen occurs in molecules consisting of two oxygen atoms. As you know, a molecule of water contains two hydrogen atoms and one oxygen atom, so the formula is H_2O.

5.2.4 PERIODIC TABLE OF ELEMENTS

One does not have to be a rocket scientist to recognize that some elements have similar chemical properties. This is especially the case if you work with chemicals on a regular basis. The point is a chemical such as bromine (atomic number 35) has chemical properties that are similar to the chemical properties of the element chlorine (atomic number 17, which most water/wastewater workers are familiar with) and iodine (atomic number 53).

The English chemist Newlands, in 1865, arranged some of the known elements in an increasing order of atomic weights. Newlands' arrangement

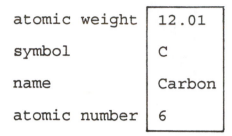

atomic weight	12.01
atomic number	
symbol	C
name	Carbon
atomic number	6

Figure 5.1 Periodic table entry for carbon.

had the lightest element he knew about at the top of his list and the heaviest element at the bottom. Newlands was surprised when he observed that starting from a given element, every eighth element repeated the properties of the given element.

Later, in 1869, Mendeleev, a Russian chemist, published a table of the sixty-three known elements. In his table Mendeleev, like Newlands, arranged the elements in an increasing order of atomic weights. He also grouped them in eight vertical columns so that elements with similar chemical properties would be found in one column. It is interesting to note that Mendeleev left blanks in his table. He correctly hypothesized that undiscovered elements existed that would fill in the blanks when they were discovered. Because he knew the chemical properties of the elements above and below the blanks in his table, he was able to predict quite accurately the properties of some of the undiscovered elements.

Today our modern form of the periodic table is based on work done by the English scientist Henry Moseley, who was killed during World War I. Following the work of Rutherford (a New Zealand physicist) and Bohr (a Danish physicist), Moseley used X-ray methods to determine the number of protons in the nucleus of an atom.

The atomic number, or number of protons, of an atom is related to its atomic structure. In turn, atomic structure governs chemical properties. The atomic number of an element is more directly related to its chemical properties than is its atomic weight. It is more logical to arrange the periodic table according to atomic numbers than atomic weights. By determining the atomic numbers of the elements, Moseley enabled chemists to make a better periodic table.

In the periodic table, each box or section contains the atomic number, symbol, and atomic weight of an element. The numbers down the left side of the box show the arrangement, or configuration, of the electrons in the various shells around the nucleus. For example, the element carbon has an atomic number of 6, its symbol is C, and its atomic weight is 12.011 (see Figure 5.1).

In the periodic table, a horizontal row of boxes is called a *period* or *series*. Hydrogen is all by itself because of its special chemical properties. Helium is the only element in the first period. The second period contains lithium, beryllium, boron, carbon, nitrogen, oxygen, fluorine, and neon. Other elements may be identified by looking at the table.

A vertical column is called a *group* or *family*. Elements in a group have similar chemical properties.

The periodic table is useful because by knowing where an element is located in the table, you can have a general idea of its chemical properties.

Water and wastewater operators routinely work with about a third of the 100+ elements listed in the periodic table. Table 5.1 shows these elements and their chemical symbols.

TABLE 5.1. Elements Water/Wastewater Workers Work With.

Element	Symbol
Aluminum	Al
Arsenic	As
Barium	Ba
Cadmium	Cd
Calcium	Ca
Carbon	C
Chlorine	Cl
Chromium	Cr
Cobalt	Co
Copper	Cu
Fluoride	F
Gold	Au
Helium	He
Hydrogen	H
Iodine	I
Iron	Fe
Lead	Pb
Magnesium	Mg
Manganese	Mn
Mercury	Hg
Nitrogen	N
Nickel	Ni
Oxygen	O
Phosphorus	P
Potassium	K
Silver	Ag
Sodium	Na
Sulfur	S
Tin	Sn
Uranium	U
Zinc	Zn

Water, H_2O, is a compound. *Compounds* are chemical substances made up of two or more elements bonded together. Unlike elements, compounds can be separated into simpler substances by chemical changes. Most forms of matter in nature are composed of combinations of the 100+ pure elements.

If you have a particle of a compound, for example a crystal of salt (sodium chloride) and subdivide, subdivide, and subdivide until you get the smallest unit of sodium chloride possible, you would have a molecule. A *molecule* (or least common denominator) is the smallest particle of a compound that still has the characteristics of that compound.

Let's refresh our memories here. Remember, the atom is defined as the smallest unit of matter retaining the characteristics of an element; a molecule is the smallest unit of matter retaining the properties of a compound.

Because the weights of atoms and molecules are relative and the units are extremely small, the chemist works with units he/she identifies as moles. A mole (symbol mol) is defined as the amount of a substance that contains as many elementary entities (atoms, molecules, and so on) as there are atoms in 12 g of the isotope carbon-12. NOTE: an isotope of an element is an atom having the same structure as the element—the same electrons orbiting the nucleus, and the same protons in the nucleus, but having more or fewer neutrons.

One mole of an element that exists as single atoms weighs as many grams as its atomic number (so one mole of carbon weighs 12 g), and it contains 6.022045×10^{23} atoms, which is *Avogadro's number.*

As stated previously, symbols are used to identify elements. This is a shorthand method for writing the names of the elements. This shorthand method is also used for writing the names of compounds. Symbols used in this manner show the kinds and numbers of different elements in the compound. These shorthand representations of chemical compounds are called chemical *formulas.* For example, the formula for table salt (sodium chloride) is NaCl. The formula shows that one atom of sodium combines with one atom of chlorine to form sodium chloride. Let's look at a little more complex formula for the compound sodium carbonate (soda ash): Na_2CO_3. The formula shows that this compound is made up of 3 elements: sodium, carbon, and oxygen. And there are two atoms of sodium, one atom of carbon, and three atoms of oxygen in each molecule.

In depicting chemical reactions, chemical *equations* are used. The following equation shows a chemical reaction that most water/wastewater workers are familiar with: chlorine gas added to water. It shows the formulas of the molecules that react together and the formulas of the product molecules.

$$Cl_2 + H_2O \rightarrow HOCl + HCl$$

As stated previously, a chemical equation tells what elements and compounds are present before and after a chemical reaction. Sulfuric acid poured over zinc will cause the release of hydrogen and the formation of zinc sulfate. This is shown by the following equation:

$$Zn + H_2SO_4 \rightarrow ZnSO_4 + H_2$$

One atom (also one molecule) of zinc unites with one molecule of sulfuric acid giving one molecule of zinc sulfate and one molecule (two atoms) of hydrogen. Notice that there is the same number of atoms of each element on each side of the arrow. However, the atoms are combined differently.

Let's look at another example.

When hydrogen gas is burned in air, the oxygen from the air unites with the hydrogen and forms water. The water is the product of burning hydrogen. This can be expressed as an equation.

$$2H_2 + O_2 \rightarrow 2H_2O$$

This equation indicates that two molecules of hydrogen unite with one molecule of oxygen to form two molecules of water.

5.2.5 IONS

An *ion* is an atom or group of atoms that carry a positive or negative charge as a result of having lost or gained one or more electrons. Ions are developed through the *ionization process*. Ionization is the formation of ions by splitting of molecules or electrolytes in solution.

Simply put, water dissolves many different molecules (the atoms making up the molecules come apart in the water—dissociation). This dissociation in water is ionization and the result is charged ions. When ionization occurs, the result is both positively (*cations*) and negatively (*anions)* charged ions.

A good example of ionization that occurs in water is when sodium chloride, table salt, is dissolved in water:

NaCl	↔	Na	+	Cl
sodium chloride		sodium ion		chloride ion
		(cation)		(anion)

There are several different ions found in water. Table 5.2 shows some of the common ions found in water.

TABLE 5.2. Some Common Ions Found in Water.

Ion	Symbol
Hydrogen	H^+
Sodium	Na^+
Potassium	K^+
Ammonium	NH_{4+}
Chloride	Cl^-
Iodide	I^-
Hypochlorite	OCl^-
Bicarbonate	HCO_{3-}
Calcium	Ca^{+2}
Iron	Fe^{+2}
Sulfide	S^{-2}
Carbonate	CO_{3-2}
Phosphate	PO_{4-3}

5.2.6 WATER'S ABILITY TO DISSOLVE

Water readily dissolves some substances more than others. For example, bases, salts, and mineral acids are easily dissolved. Organic substances (e.g., fats and oils), on the other hand, are not as easily dissolved in water.

Those substances that easily dissolve in water are affected by temperature. At a given temperature water will dissolve only so much solute (solute is the component of a solution that is dissolved by the solvent—water). This is the case because the solution becomes saturated with solute; it has reached its limit. When more solute is added to a saturated solution, it will not dissolve. It is interesting to note that when solutes are placed in water to dissolve, if the temperature is increased, the amount of solute required to reach saturation increases.

5.3 WATER: COMPONENTS OF

Water in its pure form is composed of hydrogen and oxygen, H_2O. Water, however, is *pure* only in the vapor state, and impurities begin to accumulate as soon as condensation occurs. Natural water can contain several components or substances. In water/wastewater operations these substances are called *contaminants, constituents,* or *impurities.*

As stated above, impurities begin to accumulate as soon as condensation takes place. Gases dissolve in the droplets forming clouds. Upon reaching the surface, water either percolates into the soil, becoming groundwater, or runs off along the surface in streams and rivers. Minerals contained in the earth's surface and below it dissolve in the water. Some of the chemical impurities commonly found in water include magnesium, sodium, calcium, potassium, bicarbonate, chloride, sulfate, and a wide variety of organic compounds.

Not all impurities dissolve in water. Natural water carries a lot of impurities, like large particles of silt, that will not dissolve. Water and wastewater operators are quite familiar with these impurities or solids. They are commonly referred to as *suspended solids.* Water works and wastewater treatment plants use sedimentation basins and clarifiers to allow these suspended solids to settle out so that they can be removed from the flow.

Total solids refers to the solids in water, sewage, or other liquids; it is composed of the suspended solids, which are easily removed, and other solids that are not easily removed. These other solids, *colloids*, are extremely fine suspended solids (particles) of less than one micron in diameter. Colloidal solids are difficult to remove from both water and wastewater. Bacteria and fine silt are examples of colloidal particles.

5.4 TURBIDITY

One of the first things people notice about water is its clarity. Turbidity is a condition in water caused by the presence of suspended matter, resulting in the scattering and absorption of light rays. In plain English, turbidity is a measure of the light-transmitting properties of water. Natural water that is very clear (low turbidity) allows one to see images at considerable depths. High turbidity water, on the other hand, appears cloudy. It should be pointed out that when water has low turbidity, this does not mean that it is without dissolved solids. Dissolved solids do not cause light to be scattered or absorbed; thus, the water looks clear. High turbidity causes problems for both the water and wastewater operator. In water, components that cause high turbidity can cause taste and odor problems. In wastewater, high turbidity will reduce the effectiveness of ultraviolet (UV) treatment for disinfection.

5.5 WATER COLOR

The color of water can be deceiving. Many of the colors associated with water are not true colors but the result of colloidal suspension (apparent color). This *apparent color* can be attributed to dissolved tannin extracted from decaying plant material. *True colors* are the result of dissolved chemicals, most often organics, that cannot be seen. In water treatment, color is a measure of aesthetic quality of water and has no direct health impact. In wastewater treatment, color refers to the *condition* or age of the water.

5.6 DISSOLVED OXYGEN (DO)

It was pointed out earlier that gases can be dissolved in water. Oxygen is an important gas that dissolves in water (but is only slightly soluble in water) and is important to most aquatic organisms. Additionally, dissolved oxygen (DO) is also an important indicator of water quality. In wastewater treatment, for example, the presence of DO is desirable because it helps to prevent formation of noxious odors.

It was pointed out earlier that solutions can become saturated with solute. This is the case with water and oxygen. As with other solutes, the amount of oxygen that can be dissolved at saturation depends upon temperature of the water. It is interesting to note, however, that the effect is just the opposite. The higher the temperature the lower the saturation level; the lower the temperature the higher the saturation level.

Oxygen is important to most aquatic life. Oxygen is taken in by aquatic organisms/microorganisms, utilized, and then dispelled as carbon dioxide, CO_2. This carbon dioxide is then consumed by aquatic plants and algae. Carbon dioxide is important because of the role it plays in *alkalinity* and *pH* (to be discussed later).

5.7 METALS IN WATER

Water and wastewater often carry metal impurities. Although most of the metals are not harmful at normal levels, a few metals can cause taste and odor problems in drinking water and disposal or reuse problems with wastewater. In addition, some metals may be toxic to humans, animals, and microorganisms. In water, most metals enter as part of compounds that ionize to release the metal as positive ions. In wastewater, the waste stream picks up metals mainly from industrial waste influent. This metal contamination can cause several treatment problems. For example, under EPA's 503 Regulation, biosolids (the solid waste product contained in wastewater) must be monitored for metal content. If the metal(s) is/are present and its/their content exceeds set levels, then extraordinary methods must be employed to reduce this metal content or, more importantly, the biosolids must be restricted in how it can be disposed; reuse of metal contaminated treated wastewater and/or biosolids is restricted.

5.8 ORGANIC MATTER

Organic matter or *compounds* are those that contain the element carbon and are derived from material that was once alive (i.e., plants and animals).

Organic compounds include fats, dyes, soaps, rubber products, plastics, wood, fuels, cotton, proteins, and carbohydrates. In water, organic compounds are usually large, nonpolar molecules that do not dissolve well in water. In wastewater, organic compounds are usually up to 75% of the suspended solids. Proteins, carbohydrates, fats, oils, and grease are the main organic constituents found in the waste stream. Organic surfactants, mainly soaps, and pesticides and other agricultural chemicals are also found in wastewater.

5.9 INORGANIC MATTER

Inorganic compounds are carbon-free, not derived from living matter, and easily dissolved in water; moreover, they are chemical substances of mineral origin. The inorganics include acids, bases, oxides, salts, etc. In water and wastewater, several inorganic components are important in establishing and controlling water quality. Two important inorganic constituents in water and wastewater are nitrogen and phosphorous.

5.10 pH

pH is a measure of the hydrogen ion (H^+) concentration. Solutions range from very acidic (having a high concentration of H^+ ions) to very basic (having a high concentration of OH^- ions). The pH scale ranges from 0 to 14 with 7 being the neutral value (see Figure 5.2). The pH of water or wastewater is important to the chemical reactions in both, and pH values that are too high or low can inhibit growth of microorganisms.

In regards to high and low pH values, high pH values are considered basic and low pH values are considered acidic. Stated another way, low pH values indicate a high level of H^+ concentration, while high pH values indicate a low H^+ concentration. Because of this inverse logarithmic relationship there is a tenfold difference in H^+ concentration.

Natural water varies in pH depending on its source. Pure water has a neutral pH, with an equal number of H^+ and OH^-.

$$H_2O \leftrightarrow H^+ + OH^-$$

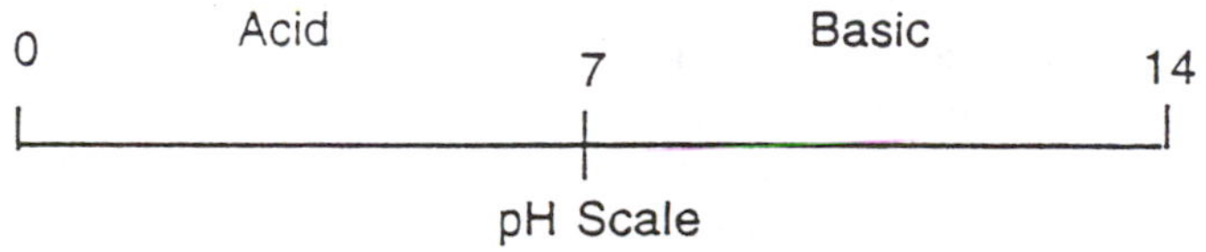

Figure 5.2 pH scale.

Adding an acid to water causes additional H^+ ions to be released so that the H^+ ion concentration goes up and the pH value goes down.

$$HCl \rightarrow H^+ + Cl^-$$

It is not only important for water and wastewater operators to know the pH of the constituents in their process or processes, it is also important to know how to make adjustments. The point is that for each water/wastewater treatment process, there is a pH at which optimum process performance can be obtained. If, for example, the pH of the water/wastewater is too low (treatment stream is too acidic) for an operation to be effective, then the pH can be increased by the addition of a base, such as calcium oxide (lime).

If the pH is too high (treatment stream is too basic), then the pH can be lowered by the addition of an acid. In water treatment, carbon dioxide is used to lower pH. In wastewater treatment, pH levels that are too high or low can have an adverse impact on biological treatment.

5.11 ALKALINITY

Alkalinity is a measure of water's ability to neutralize an acid. When acid is added to water that has a high concentration of OH^- ions (water has high pH), the H^+ ions released by the acid combine with the OH^- ions in the water to form water (H_2O). This high concentration of OH^- ions acts to neutralize acid when it is added to water. Thus, alkalinity is really a *buffer,* which tends to stabilize and prevent fluctuations in pH. In order to prevent quick changes in pH of water, it is advantageous to have significant alkalinity. Quick changes in pH interfere with the effectiveness of water/wastewater treatment processes. Another important factor about water alkalinity is that if alkalinity is low, the corrosive tendencies of water are increased. Wastewater is normally alkaline, receiving its alkalinity from groundwater, materials added during use, and from the water supply. In water, alkalinity is low when it is below 80 mg/L. In wastewater, at medium strength, alkalinity is approximately 100 mg/L.

Alkalinity should not be confused with pH. Alkalinity is the result of carbonate (CO_3^{-2}—called carbonate alkalinity), bicarbonate (HCO_3^{-2}—called bicarbonate alkalinity) and hydroxide (OH^-—called hydroxyl alkalinity) ions in the water/wastewater. Lab analysis lists the combined effect of all three as the *total alkalinity.* Typical water/wastewater chemicals used to increase alkalinity are soda ash, hydrated lime, and quick lime.

5.12 HARDNESS

In water treatment, hardness is important. *Hardness* is a characteristic of water, caused primarily by calcium and magnesium ions. Hardness can cause many maintenance problems. This is especially the case with piping and process components where scale buildup can occur. Although hardness can be present when alkalinity is low, hardness and alkalinity often occur together because some compounds contribute both alkalinity and hardness ions.

5.13 CHEMICALS USED IN WATER/WASTEWATER TREATMENT PROCESSES

In order to operate a water/wastewater treatment process successfully, you will need to know the types of chemicals used in the processes and what the purpose of each is. This section discusses chemicals and chemical unit processes used in

- odor control
- disinfection
- chemical precipitation
- adsorption
- coagulation
- taste and odor removal
- water softening
- recarbonation
- ion exchange softening
- scaling and corrosion control

5.13.1 ODOR CONTROL (WASTEWATER TREATMENT)

In wastewater treatment, odor is not a problem until the neighbors complain (Spellman, 1997a). Experience has shown that when treatment plant odor is apparent, it is not long before the neighbors do complain. The point is that odor control is an important factor affecting the performance of any wastewater treatment plant, especially in regards to public relations.

According to Metcalf & Eddy (1991), "The principal sources of odors are from (1) septic wastewater containing hydrogen sulfide and odorous compounds, (2) industrial wastes discharged into the collection system, (3) screenings and unwanted grit, (4) septage handling facilities, (5) scum on primary settling tanks, (6) organically overloaded treatment processes, (7) [biosolids]-thickening tanks, (8) waste gas-burning operations where lower-than-optimum temperatures are used, (9) [biosolids]-conditioning and dewatering facilities, (10) [biosolids] incinerator, (11) digested [biosolids] in drying beds or [biosolids]-holding basins, and (12) [biosolids]-composting operations" (p. 512).

Odor control can be accomplished by chemical or physical means. Physical means include utilizing buffer zones between the process operation and the public, making operation changes, controlling discharges to collection systems, containment, dilution, fresh air, adsorption, using activated carbon, scrubbing towers, and other means.

Odor control by chemical means involves scrubbing with various chemicals, chemical oxidation, and chemical precipitation methods. In *scrubbing with chemicals,* odorous gases are passed through specially designed scrubbing towers to remove odors. The commonly used chemical scrubbing solutions are chlorine and potassium permanganate. When hydrogen sulfide concentrations are high, sodium hydroxide is often used. In *chemical oxidation* applications, the oxidants chlorine, ozone, hydrogen peroxide, and potassium permanganate are used to oxidize the odor compounds. *Chemical precipitation* works to precipitate sulfides from odor compounds using iron and other metallic salts.

5.13.2 DISINFECTION

In water and wastewater practice, *disinfection* is often accomplished using chemicals. The purpose of disinfection is to selectively destroy disease-causing organisms. Chemicals commonly used in disinfection include chlorine and its compounds (most widely used), ozone, bromine, iodine, hydrogen peroxide, and others.

Many factors must be considered when choosing the type of chemical to be used for disinfection. These factors include: contact time, intensity and nature of the physical agent, temperature, and type and number of organisms.

5.13.3 CHEMICAL PRECIPITATION

In wastewater treatment, *chemical precipitation* is used to remove phosphorous and to enhance suspended-solids removal in sedimentation processes. The most common chemicals used are aluminum hydroxide (alum), ferric chloride, ferric sulfate, and lime.

5.13.4 ADSORPTION

In wastewater treatment, *adsorption,* using activated carbon, is utilized to remove organics not removed by biological and other chemical treatment processes. Adsorption can also be used for dechlorination of wastewater before final discharge of treated effluent.

5.13.5 COAGULATION

In water treatment, normal sedimentation processes do not always settle out particles efficiently. This is especially the case when attempting to remove particles of less than 50 μm in diameter.

In some instances it is possible to agglomerate (to make or form into a rounded mass) particles into masses or groups. These rounded masses are of increased size and therefore increased settling velocities, in some instances. For colloidal-sized particles, however, agglomeration is difficult. The point is turbid water resulting from colloidal particles is difficult to clarify without special treatment.

This special treatment process used in water treatment is *chemical coagulation*. Chemical coagulation is usually accomplished by the addition of metallic salts such as aluminum sulfate or ferric chloride. Alum is the most commonly used coagulant in water treatment and is most effective between pH ranges of 5.0 and 7.5. Sometimes polymer is added to alum to help form small flocs together for faster settling. Ferric chloride, effective down to a pH of 4.5 is sometimes used.

In addition to pH, a variety of other factors influence the chemical coagulation process, including
- temperature
- influent quality
- alkalinity
- type and amount of coagulant used
- type and length of flocculation
- type and length of mixing

5.13.6 TASTE AND ODOR REMOVAL

Although odors can be a problem with wastewater treatment, the taste and odor parameter is only associated with potable water. *Tastes* and *odors* in water may be produced by either organic or inorganic materials. The perceptions of taste and odor are closely related and often confused by water practitioners as well as by consumers. Thus, it is difficult to precisely measure either one. Experience has shown that a substance that produces an odor in water almost invariably imparts a taste as well. This is not the case, however, with taste. Taste is generally attributed to mineral substances in the water. Most of these minerals affect water taste but do not cause odors.

Along with the impact minerals can have on water taste, there are other substances or practices that can affect both water tastes and odors (e.g., metals, salts from the soil, constituents of wastewater, and end products generated from biological reactions). When water has a distinct taste but no odor, the taste might be the result of inorganic substances. Anyone who has tasted alkaline water has also tasted its biting bitterness. Then there are the salts; they not only give water that salty taste but also contribute to its bitter taste. Other than from natural causes, water can take a distinctive color or taste, or both, from human contamination of the water.

Both taste and odor in water can be produced by organic materials. Petroleum-based products are probably the prime contributors to both this taste and odor problem in water. Biological degradation or decomposition of organics in surface waters also contribute to both taste and odor problems in water. You might be familiar with that "rotten egg" taste and odor that is generated when decomposition of organics in water works to reduce sulfur. Algae is another problem. Certain species of algae produce oily substances that may result in both taste and odor. Synergy can also work to produce taste and odor problems in water. A good example of synergy affecting the taste and odor of water is when water and chlorine are mixed.

In measuring the level of odor-causing material in water, direct measurement techniques and/or quantitative testing employing the human senses is used. If the causative agents are known, *direct measurement* of materials that produce tastes and odors can be made. Moreover, for measuring taste-producing inorganics, several direct measurement analyses are available. In addition, for measuring taste- and odor-causing organics, chromatography (which is expensive and thus could be cost-prohibitive) can be used.

Qualitative tests that employ the human senses of taste and smell can also be used. One such test is known as the *threshold odor number* (TON). The TON is a standard unit of measurement and is the dilution factor required to produce a solution in which the odor is just perceptible (McGhee, 1991). In this test, differing amounts of odorous water are diluted with odor-free water to fill several 200-mL containers. Then a panel of five to ten persons is assembled to smell the mixtures and determine which container contents gives off an odor that is least detectable in comparison to the others. Then the TON of that sample is calculated, using the formula

$$\text{TON} = \frac{A + B}{A}$$

where A is the volume of odorous water (mL) and B is the volume of odor-free water required to produce a 200-mL mixture. Threshold odor numbers to various sample volumes are shown in Table 5.3.

TABLE 5.3. TON Values—Corresponding Sample Volume Diluted to 200 mL.

Sample A (volume-mL)	TON*
200	1.0
175	1.1
150	1.3
125	1.6
100	2.0
75	2.7
67	3.0
50	4.0
40	5.0
25	8.0
10	20.0
2	100.0
1	200.0

*The Public Health Service has recommended a maximum TON of 3; this value serves as a guideline rather than a legal standard (U.S. Dept. of Health, Education & Welfare, 1982).
Source: Adapted from Peavy, Rowe, & Tchobanglous (1985), p. 21.

Although the threshold odor number (TON) has been widely used in the past to express the concentration of odor-causing materials, other methods such as instrumental methods and a method called *flavor profile analysis (FPA)* (Bartels et al., 1985) are also being used. FPA is used to evaluate the sensory, including intensity, characteristics of water and to help water suppliers solve recurring taste and odor problems.

In regards to chemically treating water for odor and taste problems, oxidants such as chlorine, chlorine dioxide, ozone, and potassium permanganate can be used. These chemicals are especially effective when water is associated with an earthy or musty odor caused by the nonvolatile metabolic products of actinomycetes and blue-green algae. Tastes and odors associated with dissolved gases and some volatile organic materials are normally removed by oxygen in aeration processes.

5.13.7 WATER SOFTENING

The reduction of hardness, or *softening,* is a process commonly practiced in water treatment. *Chemical precipitation* and *ion exchange* are the two softening processes that are most commonly used. Softening of hard water is desired (for domestic users) to reduce the amount of soap used, increase the life of water heaters, and to reduce encrustation of pipes (cementing together the individual filter media grains).

In chemical precipitation, it is necessary to adjust pH. To precipitate the two ions most commonly associated with hardness in water, calcium (Ca^{+2})

and magnesium (Mg^{+2}), the pH must be raised to about 9.4 for calcium and about 10.6 for magnesium. To raise the pH to the required levels lime is added.

Chemical precipitation is accomplished by converting calcium hardness to calcium carbonate and magnesium hardness to magnesium hydroxide. This is normally accomplished by using the lime-soda ash or the caustic soda processes.

The *lime-soda ash* process reduces the total mineral content of the water, removes suspended solids, removes iron and manganese, and reduces color and bacterial numbers. The process, however, has a few disadvantages. For example, the process produces large quantities of sludge, requires careful operation, and, as stated earlier, if the pH is not properly adjusted, may create operational problems downstream of the process (McGhee, 1991).

In the *caustic soda* process, the caustic soda reacts with the alkalinity to produce carbonate ions for reduction with calcium (van der Veen & Graveland, 1988). The process works to precipitate calcium carbonate in a fluidized bed of sand grains, steel grit, marble chips, or some other similar dense material. As particles grow in size by deposition of $CaCO_3$, they migrate to the bottom of the fluidized bed from which they are removed. This process has the advantages of requiring short detention times (about 8 seconds) and producing no sludge.

5.13.8 RECARBONATION

Recarbonation (stabilization) is the adjustment of the ionic condition of a water so that it will neither corrode pipes nor deposit calcium carbonate, which produces an encrusting film. During or after the lime-soda ash softening process, this recarbonation is accomplished through the reintroduction of carbon dioxide into the water. Because lime softening of hard water supersaturates the water with calcium carbonate and may have a pH of greater than 10, pressurized carbon dioxide is bubbled into the water, lowering the pH and removing calcium carbonate.

5.13.9 ION EXCHANGE SOFTENING

Hardness can be removed by *ion exchange*. In water softening, ion exchange replaces calcium and magnesium with a nonhardness cation, usually sodium. Calcium and magnesium in solution are removed by interchange with sodium within a solids interface (matrix) through which the flow is passed. Similar to the filter, the ion exchanger contains a bed of granular material, a flow distributor, and an effluent vessel that collects the product. The exchange media include greensand (a sand or sediment given a dark greenish color by grains of glauconite), aluminum silicates, synthetic siliceous gels, bentonite clay, sulfonated coal, and synthetic organic resins and are generally in particle form usually ranging up to a diameter of 0.5 mm. Modern applications more often employ artificial organic resins. These clear, BB-sized resins are sphere-shaped and have the advantage of providing a greater number of exchange sites. Each of these resin spheres contains sodium ions, which are released into the water in exchange for calcium and magnesium. As long as exchange sites are available, the reaction is virtually instantaneous and complete.

However, when all the exchange sites have been utilized, hardness begins to appear in the influent (*breakthrough*). When breakthrough occurs, this necessitates the regeneration of the medium by contacting it with a concentrated solution of sodium chloride (Peavy et al., 1985).

Ion exchange used in water softening has both advantages and disadvantages. One of its major advantages is that it produces a softer water than does chemical precipitation. Additionally, ion exchange does not produce the large quantity of sludge encountered in the lime-soda process. There are a few disadvantages, however. For example, although it does not produce sludge, ion exchange does produce concentrated brine. Moreover, the water must be free of turbidity and particulate matter or the resin might function as a filter and become plugged.

5.13.10 SCALING AND CORROSION CONTROL

Controlling scale and corrosion is important in water systems. *Scale* is caused by carbonate and noncarbonate hardness constituents in water. It forms a chalky-white deposit frequently found on tea kettle bottoms. When controlled, this scale can be beneficial, forming a protective coating inside tanks and pipelines. A problem arises when scale is not controlled. Excessive scaling reduces the capacity of pipelines and the efficiency of heat transfer in boilers.

Corrosion is the oxidation of unprotected metal surfaces. Of particular concern, in water treatment, is the corrosion of iron and its alloys (i.e., the formation of rust). Several factors contribute to the corrosion of iron and steel. pH, alkalinity, dissolved oxygen, and carbon dioxide can all cause corrosion. Along with the corrosion potential of these chemicals, their corrosive tendencies are significantly increased when water temperature and flow are increased.

CHAPTER 6

Water Biology

Scientists picture the primordial Earth as a planet washed by a hot sea and bathed in an atmosphere containing water vapor, ammonia, methane and hydrogen. Testing this theory, Stanley Miller at the University of Chicago duplicated these conditions in the laboratory. He distilled sea water in a special apparatus, passed the vapor with ammonia, methane and hydrogen through an electrical discharge at frequent intervals, and condensed the "rain" to return to the boiling sea water. Within a week the sea water had turned red. Analysis showed that it contained amino acids, which are the building blocks of protein substances.

Whether this is what really happened early in the Earth's history is not important; the experiment demonstrated that the basic ingredients of life could have been made in some such fashion, setting the stage for life to come into existence in the sea. The saline fluids in most living things may be an inheritance from such early beginnings. (Kemmer, 1979, p. 71)

6.1 INTRODUCTION

WATER, wastewater, and other water practitioners must have considerable knowledge of the biological characteristics of water and wastewater. Water/wastewater specialists must know the principal groups of microorganisms found in water supplies (surface and groundwater) and wastewater as well as those that must be treated (pathogenic organisms) and/or controlled for biological treatment processes; they must be able to identify the organisms used as indicators of pollution and know their significance; and they must know the methods used to enumerate the indicator organisms. These matters are discussed in this chapter.

Biology is the science of life. *Microbiology* is a branch of biology that deals with the study of organisms that are mainly microscopic in size and thus cannot be readily seen except with the aid of a microscope. It is the study of basic microbiology, related to water/wastewater, that this chapter is designed to address.

117

Right about now, the reader (probably a water practitioner) might be asking him/herself whether studying microbiology is really important, necessary. In a sense, this question can be answered in a couple of ways. The first answer is more philosophical or historical rather than practical; it makes a point, however, that most would find difficult to dispute. Philosophically speaking, consider the evidence that links all life to the sea; that is, the primordial sea. The first life on earth was certainly aquatic. Exactly what this first organism looked like, we do not know. It seems certain, however, that this protocell was simple: something like the solitary blue-green algae that still inhabits warm-water springs, with traces of it in fossilized form found in Africa and Australia; some have been dated and are estimated to be 3.5+ billion years old.

Or maybe this protocell was a giant bacteria.

The fact is we do not know. We are not sure. We probably never will be sure.

The point is that some form of life began in a primordial sea; this lifeform contained amino acids (the precursors of life). From the sea, this primordial microorganism leaped, slipped, bounced, hopped, jumped, sprang, vaulted, or was blown ashore. Now, in a strange new, forbidding environment, our friend (our very good friend), the protocell, survived, took root, reproduced, flourished, and evolved.

What happened next? We do not know. It is doubtful whether we will ever know.

A better question might be: Is this really that important? It is if you are interested in "the beginning"—where we all came from. The point is that it was from our friend (our relative), the protocell, that our kind of life began. How do we know this? We know this because our bodies carry the evidence; we are a mobile reservoir carrying fluid within us that is a perfect facsimile of that sea in which we grew to realization. The fact of the matter is that the chemical makeup of our tissues and blood is exactly the same as those that once predominated in that primordial sea.

Since water is so much of our very being and since all life originated in water, from a philosophical or historical point of view, water practitioners should have some basic knowledge of the microlife present in water today.

From the practical point of view, water practitioners, specifically water and wastewater treatment plant operators, must have a basic knowledge of microbiology because treatment plant operations cannot be optimized without it. Further, most state licensing examinations test for knowledge related to microbiology and/or microbiological principles.

Let's look at each technical occupation, water and wastewater operator, and determine why knowledge about microbiology is so important.

(1) Water treatment specialists are concerned with water supply and water

purification through a treatment process. In treating water, the primary concern is producing potable water that is safe to drink (free of pathogens) with no accompanying offensive characteristics such as foul taste and odor. The treatment specialist must possess a wide range of knowledge in order to correctly examine water for pathogenic microorganisms and to determine the type of treatment necessary to ensure that the water quality of the end product, potable water, meets regulatory requirements.

(2) Wastewater treatment specialists are also concerned with water quality. However, they are not as concerned as water specialists with total removal or reduction of most microorganisms. The wastewater treatment process actually benefits from microorganisms that act to degrade organic compounds and, thus, stabilize organic matter in the waste stream. Thus, wastewater specialists must be trained to operate the treatment process in a manner that controls the growth of microorganisms and puts them to work in the stabilization process. Moreover, to more fully understand wastewater treatment, it is necessary to determine which microorganisms are present and how they function to break down the components in the wastewater stream. Then, of course, the wastewater specialist must ensure that before dumping treated effluent into a receiving body the microorganisms that worked so hard to degrade organic waste products, especially the pathogenic microorganisms, are not sent from the plant with effluent as viable organisms (Spellman, 1996b).

If you still are not convinced that having a knowledge of aquatic biology (and chemistry) is important for the water/wastewater specialist, consider this: Managers of water and wastewater treatment facilities are under increasing pressure to stretch their organizations' dollars. (It's always easier to make a point when you put the argument in terms of dollars—don't you agree?)

Are you familiar with the term *privatization?* Maybe you are or maybe you are not. The point is there is a movement, a trend, in the United States toward privatizing utilities, including water and wastewater works. The goal of privatization, of course, is to find a more efficient, less costly way in which to perform the essential tasks and functions related to water/wastewater treatment.

What privatization is really about is downsizing to save money.

When treatment managers decide how to spend money, increasing or even maintaining the size of the treatment plant staff is not a high priority—we are dispensable (no longer needed).

The key to preventing downsizing is level of qualification. The level of qualification we are talking about here, as stated previously, is the

fact that water/wastewater and other water specialists must be Jacks or
Jills of all trades. You must be valuable, irreplaceable, absolutely indispensable.

Probably the best way in which to make yourself valuable, irreplaceable,
and indispensable, in water/wastewater treatment, is to ensure that you are
not only an operator but also an analyst. A water/wastewater operator who
is also a certified lab analyst, for example, has a leg up on other operators.
When the employer looks to save money by downsizing, he/she will find it
difficult to terminate someone who is not only a well-trained operator but
who also has a basic understanding of a treatment system's biology and
chemistry.

What we are really saying here is that one should never feel comfortable
with one's present level of training and knowledge. The goal should be to
continue the educational process—to make sure that one is valuable, irreplaceable, and indispensable. For water/wastewater operators this means a
thorough knowledge of water biology and chemistry.

6.2 MICROORGANISMS

Microorganisms of interest to water/wastewater specialists include bacteria, protozoa, viruses, algae, and fungi.

Because they are a major health concern, water treatment specialists are
mostly concerned about how to control microorganisms that cause *waterborne diseases*. These waterborne diseases are carried by *waterborne
pathogens* (i.e., bacteria, virus, and protozoa).

Wastewater treatment specialists are mostly concerned about the millions of organisms that arrive at the plant with the influent. The majority
of these organisms are nonpathogenic. There are, however, pathogenic
organisms that may be present. These include the organisms responsible
for diseases such as typhoid, tetanus, hepatitis, dysentery, gastroenteritis, and others.

In order to understand how to minimize or maximize growth of microorganisms and control pathogens one must study the structure and characteristics of the microorganisms. In the sections to follow we will look at each
of the major groups of microorganisms in relation to their size, shape,
types, nutritional needs, and control. Before beginning this discussion,
however, it is important to make a point concerning waterborne disease
carried by waterborne pathogens.

In the water environment, Koren (1991) points out that water is not a medium for the growth of microorganisms, but is instead a means of transmission (a conduit for; hence, the name *waterborne*) of the pathogen to the
place where an individual is able to consume it and there start the outbreak

of disease. This is contrary to the view taken by the average person. That is, when the topic of waterborne disease is brought to his/her attention, he/she might mistakenly assume that waterborne diseases are at home in water. Nothing could be further from the truth. A water-filled ambience is not the environment in which the pathogenic organism would choose to live, that is, if it had such a choice. The point is that microorganisms do not normally grow, reproduce, languish, and thrive in watery surroundings. Pathogenic microorganisms temporarily residing in water are simply biding their time, going with the flow, waiting for their opportunity to meet up with their unsuspecting host or hosts. To a degree, when the pathogenic microorganism finds its host or hosts, it is finally home or may have found its final resting place (Spellman, 1997b).

6.2.1 MICROORGANISMS (IN GENERAL)

As stated previously, the microorganisms we are concerned with include bacteria, fungi, protozoa, algae, and viruses. These tiny organisms make up a large and diverse group of free-living forms that exist either as single cells, cell bunches, or clusters. Any and all of these organisms may be found in water and/or wastewater.

Found in abundance almost anywhere on earth, the vast majority of microorganisms are not harmful. Many microorganisms, or microbes, occur as single cells (unicellular); others are multicellular; and still others, viruses, do not have a true cellular appearance.

A single microbial cell, for the most part, exhibits the characteristic features common to other biological systems, such as metabolism, reproduction, and growth.

6.2.2 CLASSIFICATION

For centuries, scientists classified the forms of life visible to the naked eye as either animal or plant. Much of the current knowledge about living things was organized by the Swedish naturalist Carolus Linnaeus in 1735.

The importance of organizing or classifying organisms cannot be overstated, for without a classification scheme, it would be difficult to establish a criteria for identifying organisms and to arrange similar organisms into groups. Probably the most important reason for classifying organisms is to make things less confusing (Wistreich & Lechtman, 1980).

Linnaeus was quite innovative in the classification of organisms. One of his innovations is still with us today: the *binomial system of nomenclature*. Under the binomial system all organisms are generally described by a two-word scientific name, the *genus* and *species*. Genus and species are groups that are part of a hierarchy of groups of increasing size, based on their taxonomy. This hierarchy follows:

Kingdom

    Phylum

        Class

            Order

                Family

                    Genus

                        Species

Utilizing this hierarchy and Linnaeus's binomial system of nomenclature, the scientific name of any organism (as stated previously) includes both the genus and the species name. The genus name is always capitalized, while the species name begins with a lowercase letter. On occasion, when there is little chance for confusion, the genus name is abbreviated with a single capital letter. The names are always in Latin, so they are usually printed in italics or underlined. Some organisms also have English common names. Some microbe names of interest in water/wastewater treatment follow.

- *Salmonella typhi*
- *Shigella* spp.
- *Vibrio cholerae*
- *Campylobacter* spp.
- Enteropathogenic *E. coli*
- *Leptospira* spp.
- *Entamoeba histolytica*
- *Giardia lamblia*
- *Crytosporidia*

Escherichia coli is commonly known as simply *E. coli*, while *Giardia lamblia* is usually referred to by only its genus name, *Giardia*.

A simplified system of microorganism classification is used in water and wastewater. Classification is broken down into the kingdoms of animal, plant, and protista. As a general rule, the animal and plant kingdoms contain all the multicell organisms, and the protists contain all single-cell organisms. Along with microorganism classification based on the animal, plant, and protista kingdoms, microorganisms can be further classified as being *eucaryotic* or *procaryotic*. A eucaryotic organism is characterized by a cellular organization that includes a well-defined membrane. A procaryotic organism is characterized by a nucleus that lacks a limiting membrane.

6.2.3 THE CELL

Since the nineteenth century, scientists have known that all living things, whether animal or plant, are made up of cells. Moreover, the fundamental unit of all living matter, no matter how complex, is the cell. A typical cell is an entity, isolated from other cells by a membrane or cell wall. The cell membrane contains protoplasm, the living material found within it, and the nucleus. In a typical mature plant cell, the cell wall is rigid and is composed of nonliving material, while in the typical animal cell, the wall is an elastic living membrane. Cells exist in a very great variety of sizes and shapes, as well as functions. Their average size ranges from bacteria too small to be seen with the light microscope to the largest known single cell, the ostrich egg. Microbial cells also have an extensive size range, some being larger than human cells (Kordon, 1993).

6.3 BACTERIA

Bacteria are microscopic, primitive, single-celled organisms that are present in all body discharges. Bacteria range in size from 0.5–2 microns in diameter and about 1–10 microns long. Bacteria are among the most common microorganisms in water and wastewater. They may be classified in many different ways including the source of oxygen and process they use to survive (aerobic, anaerobic, anoxic, facultative); their ability to cause disease (pathogenic or nonpathogenic); their shape and many other characteristics.

Wastewater and water treatment specialists share the challenge of controlling pathogenic bacteria. Domestic wastewater normally contains huge quantities of microorganisms, such as bacteria, viruses, protozoa, and worms.

There are three general groups of bacteria based on their physical shape. Spherical-shaped bacteria are called *cocci*. Rod-shaped bacteria are called *bacilli*. Spiral-shaped bacteria (*spirilla*) make up the third group (see Figure 6.1). Within these groups there are many different arrangements. Some exist as single cells, others as pairs, as clumps of four to eight, and as chains (see Figure 6.1).

In order to survive and to multiply, most bacteria require organic food. This food comes from plant and animal material that finds its way into the water where bacteria exist. In water, bacteria use this food to produce energy and, in turn, use the energy to reproduce new cells. It should be pointed out that bacteria can use inorganics (such as iron) as an energy source.

Probably the three most important factors affecting the growth rate of

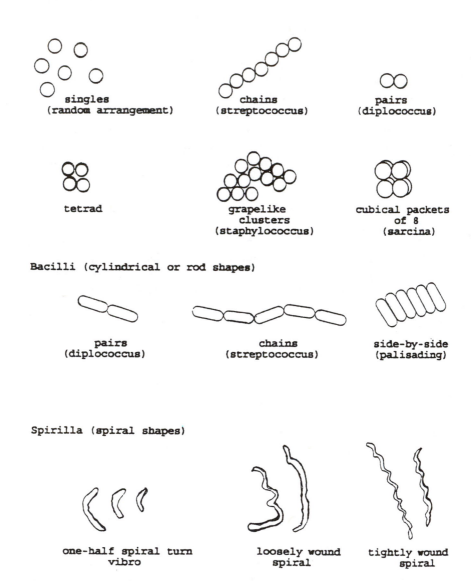

Figure 6.1 Bacterial shapes and arrangements. [Reprinted from Spellman (1997b), p. 22, with permission.]

bacteria are *temperature, pH* and *oxygen.* Although a few organisms can live in hot water (60–80°C), most prefer a moderate *temperature* (15–55°C). It is important to point out that there is a direct relationship between temperature and bacterial growth rate. The warmer the environment the faster the rate of growth. From the water/wastewater operator's point of view, this is an important relationship because, generally, for each increase of about 10°C, bacterial growth rate doubles. Obviously, this means that a bacteria contaminant will multiply more quickly when it is warm and may require more disinfection in order to obtain a "cleaner" effluent.

pH also affects bacteria growth. Most bacteria seem most comfortable in a pH range of about 6.5–8.5 (close to neutral pH). Water/wastewater that is extremely acidic or basic will inhibit bacteria growth. For example, low pH in water inhibits enzymatic activity.

Oxygen influences microscopic life. Many bacteria are *aerobic* (a condition in which dissolved oxygen is present in the aquatic environment). Aerobes require oxygen and produce carbon dioxide as a by-product. The bacteria responsible for most of the biological treatment of wastewater are aerobic. A few bacteria are *anaerobic* (an environment lacking in dissolved oxygen). They can live without oxygen. Some bacteria, called *facultative,* can grow in either aerobic or anaerobic environments. For example, in water treatment, iron bacteria is a facultative organism.

Bacteria grow rapidly when conditions are ideal. They multiply by *binary fission*, which is a simple dividing process. In binary fission, each cell separates into two identical new cells. When conditions are ideal, bacteria can double their number quickly.

In water/wastewater treatment processes, pathogenic bacteria are destroyed through *disinfection*. Disinfection does not kill off all the bacteria (that would be *sterilization*) but does reduce the number of disease-causing bacteria (and other organisms) to an acceptable level.

Even though wastewater can contain bacteria counts in the millions per ml, in wastewater treatment, under controlled conditions, bacteria can help to destroy and to identify pollutants. In such a process, bacteria stabilize organic matter, (e.g., activated sludge [biosolids] processes) and thereby assist the treatment process in producing effluent that does not impose an excessive oxygen demand on the receiving body (Kemmer, 1979). Coliform bacteria can be used as an indicator of pollution by human wastes.

6.4 PROTOZOA AND OTHER MICROORGANISMS

The *protozoa* ("first animals") are a large group of eucaryotic (possesses a clearly-defined nucleus) organisms of more than 50,000 known species that have adapted a form or cell to serve as the entire body. They are microscopic animals that are a higher life form and are normally associated with less-polluted waters. When compared in size to most other members of the microbial world, protozoa are giants. For example, they are many times larger than bacteria. They range in size from about 4 to 500 microns.

The major groups of protozoa are based on their method of locomotion (mobility or motility). Protozoa feed on organic materials, with bacteria being their favorite food. In regards to oxygen requirements, protozoa are mostly aerobic or facultative. Such factors as pH, temperature, and toxic materials affect their rate of growth.

Most protozoa have a complex life cycle in which they go from an active growth phase to an at rest stage, called *cysts*. Cysts are extremely resistant structures that protect the organism from destruction. This is why they are difficult to control in water and wastewater. Their protective cyst form may make them completely resistant to chlorine disinfection.

In water, protozoa can contribute to disease. Three common protozoan waterborne diseases are: amoebic dysentery (from *Entamoeba histolytica*), giardiasis (from *Giardia lamblia*), and cryptosporidosis (from *Cryptosporidia*).

In wastewater treatment, protozoa are a critical part of the purification process and can be used to indicate the condition of treatment processes. Protozoa normally associated with wastewater include amoeba, flagellates, free-swimming ciliates, and stalked ciliates.

Amoeba are associated with poor wastewater treatment of a young biosolids biomass (see Figure 6.2). They move through wastewater by a streaming or gliding motion. This movement is effected by moving the liquids stored within the cell wall. They are normally associated with an effluent high in BODs and suspended solids.

Flagellates (flagellated protozoa) have a single, long hair-like or whip-like projection (flagella) that is used to propel the free-swimming organisms through wastewater and to attract food (see Figure 6.3). The flagellated protozoa is normally associated with poor treatment and a young biosolids. When the predominate organism is the flagellated protozoa, the plant effluent will contain large amounts of BODs and suspended solids.

The *free swimming ciliated protozoan* uses its tiny, hair-like projections (cilia) to move itself through the wastewater and to attract food (see Figure 6.3). The free swimming ciliated protozoan is normally associated with a moderate biosolids age and effluent quality. When the free swimming ciliated protozoan is the predominate organism, the plant effluent will normally be turbid and contain a high amount of suspended solids.

The *stalked ciliated protozoan* attaches itself to the wastewater solids

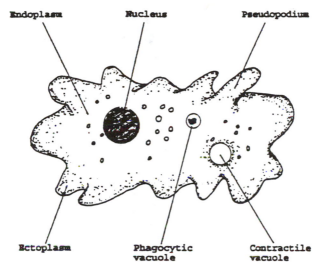

Endoplasm   **Nucleus**   **Pseudopodium**

Ectoplasm   **Phagocytic vacuole**   **Contractile vacuole**

Figure 6.2 Amoeba. [Reprinted from Spellman (1997b), p. 70, with permission.]

and uses its cilia to attract food (see Figure 6.3). The stalked ciliated protozoan is normally associated with a plant effluent that is very clear and contains low amounts of both BODs and suspended solids.

Rotifers make up a well-defined group of the smallest, simplest multicellular microorganisms and are found in nearly all aquatic habitats (see

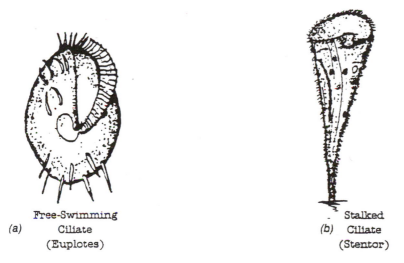

(a) Free-Swimming Ciliate (Euplotes)

(b) Stalked Ciliate (Stentor)

Figure 6.3 (a) A free-swimming ciliate; (b) a stalked ciliate. [Reprinted from Spellman (1997b), p. 69, with permission.]

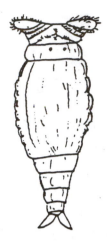

Figure 6.4 Philodina, a common rotifer. [Reprinted from Spellman (1997b), p. 75, with permission.

Figure 6.4). Rotifers are a higher life form associated with cleaner waters. Normally found in well-operated wastewater treatment plants, they can be used to indicate the performance of certain types of treatment processes.

Because they are important members of freshwater zooplankton, microscopic *crustaceans* are of interest to water and wastewater specialists. These microscopic organisms are characterized by a rigid shell structure. They are multicellular animals that are strict aerobes, and as primary producers they feed on bacteria and algae. They are important as a source of food for fish. Additionally, microscopic crustaceans have been used to clarify algae-laden effluents from oxidation ponds. *Cyclops* and *Daphnia* are the two microscopic crustaceans of interest to water and wastewater personnel (Spellman, 1997b).

Viruses are parasitic particles that are the smallest living infectious agents known. They are not cellular in that they have no nucleus, cell membrane, or cell wall. As extremely simple life forms, they multiply only within living cells (hosts) and are totally inert outside of living cells (they have no independent means of obtaining energy or to reproduce on their own) but can survive in the environment. Viruses occur in many shapes, some of which are quite complex. Because viruses lack cellular components and because they are wrapped in a relatively tough *capsid* they are not easily destroyed by normal disinfection. Usually, to effect an efficient kill, viruses must be treated with increased disinfectant concentrations and longer contact time. Waterborne diseases carried by viruses include hepatitis, poliomyelitis, and viral gastroenteritis.

6.5 ALGAE

Algae refers to a large and diverse assemblage of eucaryotic organisms that lack roots, stems, and leaves but have chlorophyll and other pigments for carrying out oxygen-producing photosynthesis. Algae are a form of aquatic plants. Moreover, they exist as microscopic, single-celled forms and also as enormous, multicellular forms (e.g., kelp). They occur in fresh water, polluted water, wastewater, and in salt water. Since most of them are photosynthetic (need sunlight), they only grow where they can be exposed to light. Thus, they grow near the surface of water.

Usually, algae are classified by their color. However, they are also commonly classified based on their cellular properties or characteristics. Several characteristics are used to classify algae including (1) cellular organization and cell wall structure; (2) the nature of the chlorophyll(s); (3) the type of motility, if any; (4) the carbon polymers that are produced and stored; and (5) the reproductive structures and methods (Spellman, 1997b).

In regards to classifying algae by their color, there are a few varieties that are of interest to freshwater and wastewater specialists. *Green algae* contain chlorophyll and are found mostly in fresh water. This form is the type that grows on basin and clarifier walls. A green-pigmented algae that resembles protozoa, the *Euglenoids* are single-celled organisms. Even though they have flagella, they are considered algae because they carry out photosynthesis. The *golden brown algae, diatoms,* are single-celled organisms that have a hard silica shell. The shells of large numbers of diatoms are mined commercially (diatomaceous earth). The *blue-green algae (Cyanobacterium)* is another algae type that undergoes photosynthesis.

For the water or wastewater treatment specialist, algae are both a nuisance and a valuable ally. Although they are not pathogenic, algae do cause problems with water treatment operations. They grow easily on the walls of troughs and basins, and heavy growth can cause plugging of intakes and screens. Additionally, algae release chemicals that often give off undesirable tastes and odors. In wastewater treatment, on the other hand, controlled algae growth can be valuable in long-term oxidation ponds where they aid in the purification process by producing oxygen.

Algae can be controlled in freshwater supplies by using chlorine and potassium permanganate. Copper sulfate is often used to control algal blooms in reservoirs.

6.6 FUNGI

Fungi are of relatively minor importance in water/wastewater operations (except for composting, where they are critical). Fungi are multicellular, autotrophic, photosynthetic protists. They grow as filamentous, mold-like forms or as yeast-like (single-celled) organisms. They feed on organic material.

6.7 BIOLOGICAL PROCESSES (WASTEWATER TREATMENT)

Uncontrolled bacteria in industrial water systems produce an endless variety of problems including disease, equipment damage, and product damage. Unlike these microbiological problems that can occur in water systems, in wastewater treatment, microbiology can be applied as a beneficial science for the destruction of pollutants in wastewater (Kemmer, 1979).

It should be noted that all the biological processes used for the treatment of wastewater (in particular) are derived from processes occurring naturally in nature. The processes discussed in the following are typical examples. It also should be noted, as Metcalf & Eddy (1991) point out, "that by controlling the environment of microorganisms, the decomposition of wastes is speeded up. Regardless of the type of waste, the biological treatment process consists of controlling the environment required for optimum growth of the microorganism involved" (p. 378).

6.7.1 AEROBIC PROCESS

In *aerobic treatment processes*, organisms use free, elemental oxygen and organic matter together with nutrients (nitrogen, phosphorus) and trace metals (iron, etc.) to produce more organisms and stable dissolved and suspended solids and carbon dioxide (see Figure 6.5).

6.7.2 ANAEROBIC PROCESS

The *anaerobic treatment process* consists of two steps, occurs completely in the absence of oxygen, and produces a useable by-product, methane gas.

In the first step of the process, facultative microorganisms use the organic matter as food to produce more organisms, volatile (organic) acids,

Oxygen		More bacteria
Bacteria	⇒	Stable solids
Organic matter		Settleable solids
Nutrients		Carbon dioxide

Figure 6.5 Aerobic decomposition.

```
Facultative Bacteria          More bacteria
Organic matter      ⇒         Volatile solids
Nutrients                     Settleable solids
                              Hydrogen Sulfide
```

Figure 6.6 Anaerobic decomposition—first step.

```
Anaerobic bacteria            More bacteria
Volatile acids      ⇒         Stable solids
Nutrients                     Settleable solids
                              Methane
```

Figure 6.7 Anaerobic decomposition—second step.

```
Nitrate oxygen                More bacteria
Bacteria                      Stable solids
Organic matter      ⇒         Settleable solids
Nutrients                     Nitrogen
```

Figure 6.8 Anoxic decomposition.

```
Sun
Algae               ⇒         More algae
Carbon dioxide                Oxygen
Nutrients
```

Figure 6.9 Photosynthesis.

carbon dioxide, hydrogen sulfide, and other gases and some stable solids (see Figure 6.6).

In the second step, anaerobic microorganisms use the volatile acids as their food source. The process produces more organisms, stable solids, and methane gas that can be used to provide energy for various treatment system components (see Figure 6.7).

6.7.3 ANOXIC PROCESS

In the *anoxic treatment process* (anoxic means without oxygen), microorganisms use the fixed oxygen in nitrate compounds as a source of energy. The process produces more organisms and removes nitrogen from the wastewater by converting it to nitrogen gas that is released into the air (see Figure 6.8).

6.7.4 PHOTOSYNTHESIS

Green algae use carbon dioxide and nutrients in the presence of sunlight and chlorophyll to produce more algae and oxygen (see Figure 6.9).

Table 6.1 lists the major biological treatment processes used in wastewater treatment, their common names, and uses.

6.8 GROWTH CYCLES

All organisms follow a basic growth cycle that can be shown as a growth curve. This curve occurs when the environmental conditions required for the particular organism are reached. It is the environmental conditions (i.e., oxygen availability, pH, temperature, presence or absence of nutrients, presence or absence of toxic materials) that determine when a particular group of organisms will predominate. Obviously, this information can be very useful in operating a biological treatment process (see Figure 6.10).

6.9 SELF-PURIFICATION

When contaminants and wastewater are discharged to a body of running water (stream or river), natural processes occur that will work to remove some forms of pollution from the water. Before industrialization, increased population growth, and massive growth in agriculture, wastes were of smaller quantities and less complex. Given sufficient time, this natural process of self-purification in running waters would remove the majority of pollutants. With increased use of more complex chemicals and materials with high levels of toxicity, the self-purification process has more difficulty.

TABLE 6.1. Major Biological Processes Used in Wastewater Treatment.

Type	Common Name	Use
Aerobic Processes:		
Suspended-growth	Activated-sludge process	Carbonaceous BOD removal
	Suspended-growth nitrification	Nitrification
	Aerated lagoons	Carbonaceous BOD removal
	Aerobic digestion	Stabilization, carbonaceous BOD removal
Attached-growth	Trickling filters	Carbonaceous BOD removal
	Roughing filters	Carbonaceous BOD removal
	Rotating biological contractors	Carbonaceous BOD removal
	Packed-bed reactors	Carbonaceous BOD removal
Combined suspended- and attached-growth processes	Activated biofilter process	Carbonaceous BOD removal
Anoxic Processes:		
Suspended-growth	Suspended-growth denitrification	Denitrification
Attached-growth	Fixed-film dentrification	Denitrification
Anaerobic Processes:		
Suspended-growth	Anaerobic digestion	Stabilization, carbonaceous BOD removal
	Anaerobic contact	Carbonaceous BOD process removal
	Upflow anaerobic sludge-blanket	Carbonaceous BOD removal
Attached-growth	Anaerobic filter process	Carbonaceous BOD removal, waste stabilization
	Expanded bed	Carbonaceous BOD removal, waste stabilization

Source: Adapted from Metcalf & Eddy (1991), pp. 380–381.

133

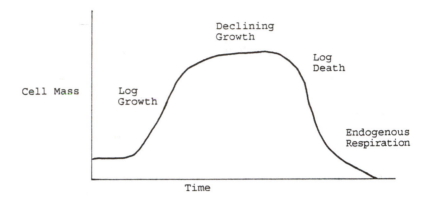

Log Growth Phase -- Organisms grow very rapidly producing large
                        numbers of new organisms.
Declining Growth -- Rate of die-off equals growth rate. Cell mass
                        growth is limited by food availability.
Endogenous Phase -- Cells must use the food accumulated in the
                        protoplasm of the cells without replenishment.

Figure 6.10 Microorganism growth curve.

To gain understanding of this self-purification process in running waters, consider the diagram shown in Figure 6.11.

As shown in Figure 6.11 the self-purification process has four zones or
stages. It should be pointed out that in the "real world" these zones are not
as clearly delineated as shown in Figure 6.11. The fact is that it is much
more difficult to distinguish when one zone ends and the next begins. The
following section describes each of these zones in greater detail.

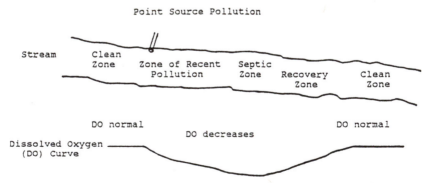

Figure 6.11 Zones of pollution and self-purification in a stream. Reprinted from Spellman
(1996b), *Stream Ecology and Self-Purification: An Introduction for Wastewater and Water
Specialists*, p. 71.

(1) Zone 1: Degradation (zone of recent pollution)
  • Wastewater enters the body of water.
  • Solids begin to settle, forming sludge banks on the bottom.
  • Dissolved oxygen levels in the stream or river decrease rapidly.
  • Water takes on the characteristic color of the wastes.
  • Fish population decreases rapidly.
  • Bacterial population increases rapidly.
(2) Zone 2: Active Decomposition (when stream aeration rate is low or non-existent—septic zone)
  • Oxygen level is zero.
  • Fish life is zero.
  • There is a high concentration of bacteria and other sewage-related organisms.
  • Color is black.
  • There is an odor of rotten eggs from hydrogen sulfide.
(3) Zone 3: Recovery Zone
  • Oxygen level begins to increase rapidly.
  • Color begins to return to normal.
  • Fish population increases.
  • Bacterial/microorganism population decreases.
  • Odor decreases.
(4) Zone 4: Clean Water Zone
  • Oxygen levels are at or near saturation.
  • Fish populations are returning to normal.
  • Color, odor return to normal.

The self-purification process removes solids that settle and organic materials that can be removed by biological activity. It will not remove toxic materials and organic matter when toxic material is present until dilution reduces the concentration of the toxic material enough to eliminate the toxic effect. Moreover, it should be noted that the self-purification process does not remove disease-causing organisms, dissolved inorganic solids, toxic materials, inorganic dyes, and others.

The self-purification process does not take place instantly. It takes time to complete. It takes time because the passage of many river miles is involved. Even with enough time and many river miles travelled, degradation or decomposition occurs anytime more wastes are added to the running water body.

6.10 BIOGEOCHEMICAL CYCLES

Several chemicals are essential to life and follow predictable cycles through nature. In these natural cycles or *biogeochemical cycles*, the chemicals are converted from one form to another as they progress through the environment. The water/wastewater plant operator should be aware of these cycles since they have a major impact on the performance of the plant and may require changes in operation at various times of the year to keep them functioning properly; this is especially the case in wastewater treatment.

Smith (1974) categorizes biogeochemical cycles into two types, the *gaseous* and the *sedimentary*. Gaseous cycles include the carbon and nitrogen cycles. The main sink of nutrients in the gaseous cycle is the atmosphere and the ocean. Sedimentary cycles include the sulfur cycle. The main sink for sedimentary cycles is soil and rocks of the earth's crust.

6.10.1 CARBON CYCLE

Carbon, which is an essential ingredient of all living things, is the basic building block of the large organic molecules necessary for life. Carbon is cycled into food chains from the atmosphere, as shown in Figure 6.12.

From Figure 6.12 it can be seen that green plants obtain carbon dioxide (CO_2) from the air and, through photosynthesis, described by Asimov

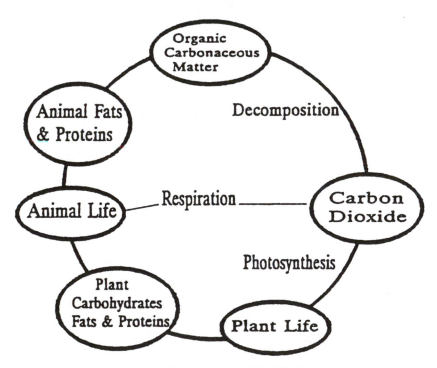

Figure 6.12 Carbon cycle (simplified).

(1989) as the "most important chemical process on Earth" (p. 20), it produces the food and oxygen that all organisms live on. Part of the carbon produced remains in living matter, the other part is released as CO_2 in cellular respiration. Miller (1988) points out that the carbon dioxide released by cellular respiration in all living organisms is returned to the atmosphere.

Some carbon is contained in buried dead animal and plant materials. Moran, Morgan, and Wiersma (1986) note that much of these buried plant and animal materials were transformed into fossil fuels. Fossil fuels, coal, oil, and natural gas, contain large amounts of carbon. When fossil fuels are burned, stored carbon combines with oxygen in the air to form carbon dioxide, which enters the atmosphere.

In the atmosphere, carbon dioxide acts as a beneficial heat screen as it does not allow the radiation of earth's heat into space. This balance is important. The problem is that as more carbon dioxide from burning is released into the atmosphere, the balance can and is being altered. Odum (1983) warns that the recent increase in consumption of fossil fuels "coupled with the decrease in 'removal capacity' of the green belt is beginning to exceed the delicate balance" (p. 202). Massive increases of carbon dioxide into the atmosphere tend to increase the possibility of global warming. The consequences of global warming "would be catastrophic . . . and the resulting climatic change would be irreversible" (Abrahamson, 1988, p.4).

6.10.2 NITROGEN CYCLE

Nitrogen is an essential element that all organisms need. In animals, nitrogen is a component of crucial organic molecules such as proteins and DNA and constitutes 1–3% dry weight of cells. Our atmosphere contains 78% by volume of nitrogen, yet it is not a common element on earth. Although nitrogen is an essential ingredient for plant growth, it is chemically very inactive, and before it can be incorporated by the vast majority of the biomass, it must be *fixed*. Nitrogen is fixed by special nitrogen-fixing bacteria found in soil and water. These bacteria have the ability to take nitrogen gas from the air and convert it to nitrate. This is called *nitrogen fixation*. Some of these bacteria occur as free-living organisms in the soil. Others live in a *symbiotic relationship* (a close relationship between two organisms of different species, and one where both partners benefit from the association) with plants. An example of a symbiotic relationship, related to nitrogen, can be seen, for example, in the roots of peas. These roots have small swellings along their length. These contain millions of symbiotic bacteria, which have the ability to take nitrogen gas from the atmosphere and convert it to nitrates that can be used by the plant. Then the plant is plowed into the soil after the growing season to improve the nitrogen content. Price (1984) describes the nitrogen cycle as an example "of a

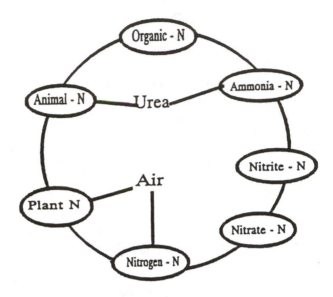

Figure 6.13 Nitrogen cycle (simplified).

largely compete chemical cycle in ecosystems with little leaching out of the system" (p. 11). Simply stated, the nitrogen cycle provides various bridges between the atmospheric reservoirs and the biological communities (see Figure 6.13).

Atmospheric nitrogen is fixed either by natural or industrial means. Moreover, nitrogen is fixed by lightning or by soil bacteria that convert it to ammonia, then to nitrite, and finally to nitrates, which can be used by plants. Nitrifying bacteria make nitrogen from animal wastes. Denitrifying bacteria convert nitrates back to nitrogen and release it as nitrogen gas.

The question becomes: What does all this have to do with water? Good question.

The best way to answer this question is to ask another question. Have you ever had the noxious pleasure of diving into a slow-moving stream covered with algal bloom? Of course, in order to have done this, you probably had to wear some type of nose plug—otherwise the horrendous stink might have driven you to pursue other activities.

If too much nitrate, for example, enters the water supply—as runoff from fertilizers—it produces an overabundance of algae, called algal bloom. The point is if this runoff from fertilizer gets into a body of water, algae may grow so profusely that they form a blanket over the surface. This usually happens in summer, when the light levels and warm temperatures favor rapid growth.

Metcalf & Eddy, Inc. (1991), in their voluminous and authoritative text, *Wastewater Engineering: Treatment, Disposal, & Reuse,* point out that nitrogen is found in wastewater in the form of urea. During wastewater treatment, the urea is transformed into ammonia nitrogen. Since ammonia exerts a BOD and chlorine demand, high quantities of ammonia in wastewater effluents are undesirable. The process of nitrification is utilized to convert ammonia to nitrates. *Nitrification* is a biological process, that involves the addition of oxygen to the wastewater. If further treatment is necessary, another biological process called *denitrification* is used. In this process, nitrate is converted into nitrogen gas, which is lost to the atmosphere, as can be seen in Figure 6.13. From the wastewater operator's point of view, nitrogen and phosphorus are both considered limiting factors for productivity. Of the two, nitrogen is harder to control, but is found in smaller quantities in wastewater.

6.10.3 SULFUR CYCLE

Sulfur, like nitrogen, is characteristic of organic compounds. The *sulfur cycle* is both sedimentary and gaseous. Tchobanglous and Schroeder (1985) note that "the principal forms of sulfur that are of special significance in water quality management are organic sulfur, hydrogen sulfide, elemental sulfur and sulfate" (p. 184).

Bacteria play a major role in the conversion of sulfur from one form to another. In an anaerobic environment, bacteria break down organic matter thereby producing hydrogen sulfide with its characteristic rotten-egg odor. A bacteria called *Beggiatoa* converts hydrogen sulfide into elemental sulfur into sulfates. Other sulfates are contributed by the dissolving of rocks and some sulfur dioxide. Sulfur is incorporated by plants into proteins. Some of these plants are then consumed by organisms. Sulfur from proteins is liberated by many heterotrophic anaerobic bacteria, as hydrogen sulfide.

6.11 REQUIREMENTS FOR BIOLOGICAL ACTIVITY

In order to have biological activity the body of water or wastewater treatment plant must possess the appropriate environmental conditions. The majority of wastewater treatment processes, for example, are designed to operate using an aerobic process. The conditions required for aerobic operation are (1) sufficient free, elemental oxygen, (2) sufficient organic matter (food), (3) sufficient water, (4) enough nitrogen and phosphorus (nutrients) to permit oxidation of the available carbon materials, (5) proper pH (6.5 to 9.0), and (6) lack of toxic materials.

CHAPTER 7

Water Ecology

Man has been interested in ecology in a practical sort of way since early in his history. In primitive society every individual, to survive, needed to have definite knowledge of his environment, i.e., of the forces of nature and of the plants and animals around him. Civilization, in fact, began when man learned to use fire and other tools to modify his environment. It is even more necessary than ever for mankind as a whole to have an intelligent knowledge of the environment if our complex civilization is to survive, since the basic "laws of nature" have not been repealed; only their complexion and quantitative relations have changed, as the world's human population has increased and as man's power to alter the environment has expanded. (Odum, 1971, p. vii)

7.1 INTRODUCTION

WHEN we think about a stream or river or some other body of water, we have two general views or examples to chose from. Review the two examples presented in the following and determine which one you prefer, which one you detest (and hopefully reject and hope to prevent).

Example 1

Whether they come from the melt snow or from a mountain spring, the headwaters of small springs are wonderfully cool. Their pools are open invitations to lie down and drink, and to peer into the tangle of delicately colored algae and mosses beneath the clear surface. Smooth, rocky shallows and riffles alternate with deeper areas where the water dams briefly against a glacial boulder and hardly seems to move at all. But most such streams are skittish in their youth, and the major problem facing life with them is to find some way of holding on against the flow.

Plants tend to develop long and slippery streamers that float in the current like windblown hair. And animals, mainly the larvae of stone and caddis flies, grow sets of hooks and grapples to anchor themselves in the current.

Downstream, things slow up a bit. The waters widen into pools overhung by sedge and grass, surrounded often by willow or alder. These shady basins are the haunt of salamanders and snails, and skitter with a profusion of tiny crustaceans, water fleas, and brightly colored water mites. They are, as a result, attractive too for frogs and snakes, and to brookside birds and animals, the dippers and thrushes, the voles and water rats, that come to feed upon them. (Watson, 1988, pp. 97, 99)

Example 2

An excursion to the local stream can be a relaxing and enjoyable undertaking. On the other hand, when you arrive at the local stream, spread your blanket on the streambank and then look out upon the stream's flowing mass only to discover a parade of waste and discarded rubble bobbing along the stream's course and cluttering the adjacent shoreline and downstream areas, any feeling of relaxation or enjoyment is quickly extinguished. Further, the sickening sensation the observer feels is not lessened, but made worse as he gains closer scrutiny of the putrid flow. He easily recognizes the rainbow-colored shimmer of an oil slick, interrupted here and there by dead fish and floating refuse, and the slimy fungal growth that prevails. At the same time, the observer's sense of smell is alerted to the noxious conditions. Along with the fouled water and the stench of rot-filled air, the observer notices the ultimate insult and tragedy: The signs warn, "DANGER—NO SWIMMING or FISHING." The observer soon realizes that the stream before him is not a stream at all; it is little more than an unsightly drainage ditch. The observer has discovered what ecologists have known and warned about for years. That is, contrary to popular belief, rivers and streams do not have an infinite capacity for pollution. (Spellman, 1996b, p. 65)

You may have experienced the conditions described in Example 1, Example 2, or both. Let's hope that your experience has only been of the type described in Example 1. Unfortunately, this is probably not the case. The point is it is not that uncommon for all of us to encounter the conditions presented in Example 2. This is a shame, of course. More pointedly, it is a tragedy. It is a tragedy of our own making. However, it is not too late to change the conditions in Example 2 back to the natural conditions described in Example 1.

7.2 WHY STUDY WATER ECOLOGY?

Along with the conditions pointed out in the preceding examples, there are other reasons for the water/wastewater specialist to study water ecology. However, in order to gain understanding of these reasons, it is necessary to gain understanding of ecology; that is, what is ecology?

Probably the best way to "gain a feel" for what ecology is and for what it is all about is to review the following:

> We poison the caddis flies in a stream and the salmon runs dwindle and die. We poison the gnats in a lake and the poison travels from link to link of the food chain and soon the birds of the lake margins become victims. We spray our elms and the following springs are silent of robin song, not because we sprayed the robins directly but because the poison traveled, step by step, through the now familiar elm leaf-earthworm-robin cycle. These are matters of record, observable, part of the visible world around us. They reflect the web of life—or death—that scientists know as ecology. (Carson, 1962, p. 189)

As Rachel Carson points out, what we do to any part of our environment has an impact upon other parts. There is an interrelationship between the parts that make up the environment. *Ecology*, simply defined, is the study of organisms at home. More specifically, ecology is the study of an organism at its home. And the home we are talking about in this text, of course, is water.

7.2.1 CASE STUDY

It was not that long ago when a young wastewater treatment plant operator assistant, who was serving an apprenticeship to become a full-fledged operator and eligible to sit for licensing examinations, was assigned the daunting task of clearing brush from the plant's stream, its outfall. Normally, this particular assistant looked forward to this type of outdoor activity. Yes, indeed, it's always great to be outside in the sunshine and fresh air on a warm spring day.

This was the attitude this young person had on this particular day. At least, this is how she felt when her day started. It wasn't all that long, however, before the young assistant came back to the plant office and complained to anyone who would listen about the "nasty flies" that were "everywhere" around the stream.

"What kind of flies you talkin' about, Sarah?" was the question the lab specialist asked.

Sarah described the flies.

The lab specialist seemed keenly interested in her descriptions. While Sarah described these "nasty bugs," the lab specialist pulled a book from the shelf. She opened the text to the section that described aquatic insects. Next to each description was a picture of each insect type.

While the assistant described the various insects that had surrounded her, pestered her, and disgusted her while working down at the stream, the lab specialist penciled a check mark next to the picture of several insects the assistant mentioned.

When the assistant finished her descriptions, the lab specialist smiled and then explained to the assistant that although the insects that she had encountered while clearing brush from the stream were an annoyance, the alternative to her discovery would be much more annoying. This would especially be the case to the plant manager who was responsible for treating the wastewater and producing an effluent for deposit into the stream that was cleaner than the stream itself.

The lab specialist went on to explain (to provide a lesson in basic stream ecology) that the dragon flies, caddis flies, mayflies, and stone flies the assistant had encountered were harbingers of good news. More specifically (and importantly) these insects are indicators. They are indicators that normally provide ecologists with an instant assessment tool for judging the water quality of a surface water body. Thus, the fact that the stream is home to dragon flies, caddis flies, and mayflies is good news; they indicate that the stream is healthy, not polluted.

The lab specialist went on with her discussion. She pointed out that the alternative to the insects that the assistant had encountered could have been quite different. For example, this particular stream might have been devoid of flies or at least devoid of the indicators of healthy conditions in the stream. In their place this stream could have contained organisms such as rat-tailed maggots, bloodworm, and tubifex worms.

The final question the lab specialist asked the assistant was: "Would you rather put up with dragon flies or rat-tailed maggots?"

The assistant answered that she preferred neither. But she understood the significance. She understood the lab specialist's point. She understood that her plant was treating wastewater correctly. She understood that maintaining a healthy stream is important. More importantly, she understood an important lesson in ecology.

7.3 GENERAL ASPECTS OF RUNNING WATERS

The diversity of ecology is enormously broad. With this said, keep in mind that this text deals with the study of the science of water and its major component parts, concepts, and applications. Thus, the major component part we are dealing with in this section is narrowly focused on stream or river ecology; the ecology of running waters. This focus is not only the obvious direction this text should take but also makes sense for a more important reason. That is, water treatment specialists take in surface water for treatment mostly from a source of running water. Further, wastewater treatment plant operators discharge plant effluent into some type of running water body. The point is that in order to round out their overall knowledge related to their water specialty, it logically follows that some knowledge of the ecology of running waters is important.

Accepting the premise that the diversity of ecology is enormously broad, it should be pointed out that the diversity of running water environments is also enormously broad. When one considers the giant leaps that have been made in new understanding of ecological, hydrological, and geomorphological processes, it is apparent that this new information works to widen the diversity of the field. At the same time, we must also consider the stream or river itself. Each is different; each is location-specific. This is not too difficult to understand and to accept when one considers that each location-specific stream is unique, impacted by local environmental factors. Although the setting of each stream contributes to its uniqueness, it should be pointed out that the processes acting in streams and rivers are general. It is these general ecological aspects of running waters that this text addresses.

Before moving on to the general ecological aspects of running waters, it should be pointed out that no text, let alone a single chapter, can cover all aspects of running water ecology. When one considers the diversity of location, the diversity of taxa, and the range of general ecological aspects, it becomes clear that the overall range of topics is too great.

However, if the reader detects a bias aimed toward discussion of biotic indicators of stream health in this text, he or she is right on target.

7.3.1 RUNNING WATERS: VARIATIONS FROM PLACE TO PLACE

In order to form a body of running water, river or stream, several components are required. In the first place, water must be available. Although running waters transport only a tiny fraction of earth's freshwater to the earth's oceans, they are an important part of the hydrologic cycle; these flowing masses of water are not insignificant. The question becomes: Where do these running waters get their water? Obviously, some water is contributed by rainfall. Some water begins the flow of a stream or enters a running water body from groundwater (springs). Water inputs are also received from snow melt-off. Additional inputs are received from melting glacial ice. A smaller amount, except in the case of intense storm events, enters streams by overland flow. Although obvious, the point is running waters need a *source* or sources of water.

Along with sources of water, running water must have a conduit in which the water can be transported or conveyed. Normally, catchment basins, river channels, and tributaries form the conduit in which water is transported.

This running water conduit has certain characteristics that are universal from stream to stream. Running water may begin as a trickle. This trickle soon develops into an increasing mass of fluid motion. This is the case due to groundwater and tributary inputs. Above the floodplain, running water flow, at some point, increases in velocity with a corresponding increase in depth and little increase in width. It should be pointed out, however, that width can increase exponentially during flood conditions (this is especially the case within the floodplain). As one proceeds downstream a body of running water enlarges, and depth and velocity also increase. The increase of flow with width is greater than the increase of depth, while velocity increases least with discharge and may remain almost constant.

From source to mouth, running waters are dynamic; they are constantly changing. When the body of running water originates in a mountainous area as a series of springs and rivulets into a fast-flowing, turbulent stream, continued downward flow and addition of tributaries result into a metamorphosis: from a crashing, smashing, turbulent, foaming, get-out-of-my-way flow to a large and smoothly sinuous river that winds its quiet way through the lowlands to the sea.

Along with the obvious change that occurs from steep slopes of the headwaters to less so downstream, materials contained within the body of running water change. In upland areas, boulders, gravel, and other coarse materials are typical. In large lowland areas, the substrate consists of softer, finer materials.

Another dynamic characteristic of running waters is also apparent in the constituents that are contained in them. While mountain streams appear pure, they are quickly transformed into muddy flows when flow receives input from farms. Moreover, flows receive rich ingredients from limestone-rich regions and almost none from flow through granite rock beds. When water flows through heavily populated areas, the heavy-hand-of-man pollutes the flow. The point is that many factors influence the composition of running waters, causing variations from place to place.

Another variation in running waters is apparent when comparing chemical composition. Variation from place to place is determined by many factors such as the amount and composition of rain and the types of rocks available for weathering. The constituents of running water include suspended inorganic matter, dissolved major ions, suspended and dissolved organic matter, dissolved nutrients, gases, and trace metals. The gases of interest, of course, are oxygen and carbon dioxide. The concentration of both is maintained close to equilibrium through exchange with the atmosphere. Further, this equilibrium is influenced by temperature and atmospheric pressure. In running waters containing macrophytes (flowering plants, mosses, liverworts, and some lichens) and filamentous algae, photosynthesis can elevate oxygen content to saturated levels that fluctuate between day and night. An opposite effect is brought about by respiration where oxygen is reduced and CO_2 elevated. Organic waste also has an impact. This is especially the case when the level of organics is high, which reduce oxygen concentration below life-sustaining levels, and elevate CO_2 to extremely high levels.

7.3.2 MAJOR PHYSICAL FACTORS OF RUNNING WATERS

Current, substrate, temperature, and *oxygen content* are the major physical factors of importance in running waters. These factors make up the physical environment that poses special challenges to the organisms that dwell there.

Not surprisingly, *current* (or directional flow) is the defining factor that characterizes all running waters and distinguishes them from other aquatic environments. Research on the anatomical and behavioral adaptations of organisms suited to survive in strong currents provides ample evidence that current is of direct importance to the life forms (biota) of these environments.

It is interesting to note that when one looks out upon the swift flow of a turbulent stream or river, the moving mass of liquid flow before you is deceptive. You see what you see but you do not see the entire picture, related to flow. This is the case because the velocity of the flow is not uniform from surface to bottom (substrate). Water in contact with many substrates (non-eroding types) has zero velocity (this region of reduced flow permits specially adapted benthic organisms to avoid fluid forces). Just above the substrate, where friction from contact with the bottom and sides comes into play, current velocity is low. At or near the surface, where friction no longer affects streamflow, flow velocity is highest.

Substrate is highly heterogeneous, consisting of both organic and inorganic materials. However, whether organic or inorganic, substrate has numerous effects, directly and indirectly, on the biota of running waters. Directly, substrate provides a surface to cling to or burrow in, a refuge from predators, material for construction of cases and tubes (a sort of home that also serves as body armor), and shelter from current. Indirectly, various other factors, such as the supply of oxygen and nutrients, supply of water, and the availability of food, all interact with substrate.

Substrates vary in type based on composition. Some are hard or soft; some are organic, others are inorganic. The point is that the type of substrate present in a body of running water has a bearing on the taxa of biota present.

Running water exhibits a wide range of temperature extremes, including some that are constant and others that vary substantially from headwaters to river mouth and on a daily and seasonal basis. Some organisms tolerate a wider temperature range than others.

The availability and the demand for *oxygen* is influenced by temperature. Although not always a limiting factor to biological activity in running waters, when flow is low and temperature is high, oxygen can be a critical variable.

7.4 LOTIC HABITAT

According to the American Society for Testing and Materials (ASTM, 1969), normal stream life can be compared to that of a *balanced aquarium* (p. 86). Nature continuously strives to provide clean, healthy, normal running waters. This is accomplished by maintaining the running water's flora and fauna in a balanced state. Nature balances stream life by maintaining both the number and the types of species present in any one part of the stream. Such balance ensures that there is never an overabundance of one species compared to another. In running waters, nature structures the stream environment so that both plant and animal life is dependent upon the existence of others within the stream. Thus, nature has structured an environment that provides for interdependence, which leads to a *balanced aquarium* in a normal body of running water (Spellman, 1996b).

As pointed out earlier, running waters or *lotic* habitats are characterized by current flow. These running water bodies, rivers and streams, have typically two zones: *riffle* (rapids) and *pool*. The *riffle* zone is the region where the velocity of current is great enough to keep the bottom clear of silt and sludge, thus providing a firm bottom for organisms. This zone contains specialized organisms that are adapted to live in running water. For example, organisms adapted to live in riffles (trout) have streamlined bodies, which aid in their ability to obtain food and in respiration. Running water organisms that live under rocks to avoid the strong current have streamlined or flat bodies. Others have suckers or hooks to cling or attach to a firm substrate to avoid the washing-away effect of the strong current (Odum, 1971).

The *pool zone* of a running water body is usually a deeper water region where velocity of water is reduced and silt and other settling solids provide a soft bottom, which is unfavorable for sensitive burrowing bottom dwellers. It is interesting to note that some stream organisms (e.g., trout) spend part of their time in the riffle part of the stream and other times in the pool zone.

Organisms are sometimes classified according to their mode of life. The following section provides a listing of the various classifications based on mode of life.

7.4.1 CLASSIFICATION OF AQUATIC ORGANISMS BASED ON MODE OF LIFE[1]

(1) *Benthos* (Mud Dwellers): The term originates from the Greek word for bottom and broadly includes aquatic organisms living on the bottom or

[1]Sections 7.4.1 and 7.4.2 are taken with permission from *Stream Ecology & Self-Purification: An Introduction for Wastewater and Water Specialists*, 1996, Technomic Publishing Co., Inc. pp. 43–45.

on submerged vegetation (Wetzel, 1983).They live under and on rocks and in sediments. A shallow sandy bottom has sponges, snails, earthworms, and some insects. A deep, muddy bottom will support clams, crayfish, and nymphs of damselflies, dragonflies, and mayflies. A firm, shallow, rocky bottom has nymphs of mayflies, stone flies and larvae of water beetles (Hickman, Roberts, and Hickman, 1988).

(2) *Periphytons* or *Aufwucks*: The first term usually refers to microfloral growth upon substrata. The second term, *Aufwucks* (German: growth upon), refers to organisms that attach or cling to stems and leaves of rooted plants or other objects projecting above the bottom without penetrating the surface (Wetzel, 1983). Insects such as mayflies, stone flies and some beetles belong to this group.

(3) *Planktons* (Drifters): They are small, mostly microscopic plants and animals that are unable to navigate against the current. They mostly float in the direction of current. There are two types of planktons: (a) Phytoplanktons are assemblages of small plants (algae) and have limited locomotion abilities; they are subject to movement and distribution by water movements. (b) Zooplanktons are animals that are suspended in water and have limited means of locomotion. Examples of zooplanktons include crustaceans, protozoans, and rotifers.

(4) *Nektons* or *Pelagic Organisms* (capable of living in open waters): They are distinct from other planktons in that they are capable of swimming independently of turbulence. They are swimmers that can navigate against the current. Examples of nektons include fish, turtles, birds, and large crayfish.

(5) *Neustons:* They are organisms that float or rest on the surface of water. Some varieties can spread out their legs so that the surface tension of the water is not broken; for example, water striders.

7.4.2 LIMITING FACTORS

A limiting factor is a condition or a substance (the resource in shortest supply) that limits the presence of an organism or a group of organisms in an area (Price, 1984). There are two well-known laws about limiting factors:

(1) *Liebigs's Law of the Minimum*: Odum (1971) has modernized Liebig's Law in the following: "Under *steady state* conditions the essential material available in amounts most closely approaching the critical minimum needed, will tend to be the limiting one" (p. 106). Liebig's Law is normally restricted to chemicals that limit plant growth in the soil, for instance nitrogen, phosphorus, and potassium. It does not deal with the excess of a factor as limiting.

(2) *Shelford's Law of Tolerance*: Although Liebig's Law does not deal with the excess of a factor as limiting, excess is or can be a limiting factor (Price, 1984). The presence and success of an organism depends on the completeness of a complex of conditions. Odum (1971) describes Shelford's Law of Tolerance as follows: "Absences or failure of an organism can be controlled by the qualitative and quantitative deficiency or excess with respect to any one of the several factors which may approach the limits of tolerance for that organism" (p. 107). For instance, too much or too little heat, light, and moisture can be limiting factors for some plants.

Price (1984) points out that "these two laws actually relate to individual organisms, and the survival of an individual in a given set of conditions, independent of others in the same niche" (p. 415). Expressed differently, both laws state that the presence and success of an organism or a group of organisms depend upon a complex of conditions, and any condition that approaches or exceeds the limits of tolerance is said to be a limiting condition or factor.

Common physical limiting factors in freshwater ecology include:

(1) Temperature
(2) Light
(3) Turbidity
(4) Dissolved atmospheric gases, especially oxygen
(5) Biogenic salts in macro and micronutrient forms
  • macronutrients such as nitrogen, phosphorus, potassium, calcium, and sulfur
  • micronutrients such as iron, copper, zinc, chlorine, and sodium
(6) Water movement—stream currents, especially rapids

7.5 SOURCES OF ENERGY TO RUNNING WATER FOOD WEBS

Organisms that can synthesize organic molecules from inorganic sources using sunlight or another source of energy are *autotrophs* (Gr. *autos*, self + *trophs,* feeder), whereas organisms that depend on organic molecules they have not synthesized for their food supplies are known as *heterotrophs* (*heteros,* another + *trophs,* feeder). In running waters, the important autotrophs are green plants, some bacteria, and protists. *Macrophytes* (large plants), *periphyton* (attached to substrates), and *phytoplankton* (suspended in water) are the major autotrophs of running waters. Heterotrophs consist of all other animals, including fungi and bacteria that gain nourishment from dead organic matter.

Macrophytes, periphytons, and river phytoplankton, as the brief description above indicates, constitute three very different groups of autotrophs occurring in running waters. Macrophytes are most abundant in small-sized rivers and backwaters, but also are found along the margins of larger rivers. Periphytons are found attached to surfaces and substrates of streams and small rivers. Populations of phytoplankton like to develop in large lowland rivers where flow is moderate. It should be pointed out that river size and condition affects the relative importance of these primary producers.

As stated above, macrophytes are found in flowing water such as canals and river margins where neither current or depth are great. There are four major categories of macrophytes: *emergent* (found on banks and shoals), *floating-leaved types* (on margins of slow rivers), *free-floating plants* (usually occur in mats), and *submerged types* attached to substrate. Limiting factors include the length of the growing season, current, and light.

Periphytons such as green algae, diatoms, blue-green algae, and a few other groups, take up residence on virtually every surface in running waters including stones, sediments, and macrophytes. Multiple factors vary their abundance and spatial distribution. These factors include high velocity flows, dimmed sunlight due to dense forest canopies, short supplies of critical nutrients such as phosphorous, and predation (grazing) by a wide variety of herbivores.

The phytoplankton are restricted primarily to lowland river regions. In order to become self-sustaining, phytoplankton must have sufficient residence time in one favorable location. Probably the most significant limiting factor affecting phytoplankton populations is sunlight. Even in those areas where sunlight is unrestricted, little light is able to penetrate to depths sufficient enough to encourage growth, especially in turbid waters. This is the main reason that phytoplankton population in rivers is not as dense as it may be in standing water.

In running waters, nonliving particulate organic matter and dissolved organic matter are important energy sources; they comprise a diverse array of potential food sources for consumers. The majority of particulate organic matter and dissolved organic matter is provided to the running water body from external inputs.

An important source of particulate organic matter is from leaves. In autumn, during leaf-fall, leaves that find their way into streams begin to break down to other constituents. For example, the chemical constituents of leaves are leeched out, then fungi begin to colonize, invertebrates feed, and physical abrasion all work to breakdown leaves to finer particulate organic matter.

Leaves are only one source of fine particulate organic matter in running waters. For example, in small woodland streams, soil and materials from the forest floor are contributors. It is important to keep in mind that the very fine particles from these sources are the end products of excessive processing and are usually low in nutritional value. The fine particles that are of most value to consumers consist of animal feces, fresh leaf fragments, and algal cells.

Dissolved organic matter also has many sources, mostly coming from outside the running water body. Leachate from fresh leaves, simple sugars, and low molecular compounds are sources that are taken up rapidly.

Microbes attached to substrates, bacteria, algae, fungi, protozoa, and detrital particles are the primary sources of uptake in small streams.

Running water ecosystems obtain their energy from both instream primary producers (plants) and from instream and externally produced non-living matter. The point is that this heterogenous mixture of autotrophs and heterotrophs provides energy and thus support for higher trophic level organisms.

While in running waters, stream-dwelling organisms, diatoms, larval vertebrates, and many different invertebrates, are not necessarily limited or locked in one location. After all, the distinguishing feature of running water is flow. Thus, stream-dwelling organisms often are transported downstream in large numbers. This large downstream movement of organisms is known as *drift*.

7.6 LOTIC ECOSYSTEMS

Most lotic (running water) ecosystems have several unique characteristics. For example, in streams, temperature and dissolved oxygen (DO) are generally evenly distributed. There may be some variation in DO between rapidly flowing, turbulent areas and deeper, quiet pools of a stream due a lacking of physical aeration in the stream. The amount of oxygen in aquatic systems is controlled by the solubility of gaseous oxygen in water. Since the DO is usually high and evenly distributed (the amount of DO in streams is generally eight to ten parts per million, ppm—ppm is like a thimble full of water inside a hot tub), the stream organisms are adapted to this environment and have a narrow range of tolerance for DO. As stated above, the exact amount of oxygen in an aquatic system is controlled by the solubility of gaseous oxygen in water. This DO content has a direct bearing on the types of organisms living in the stream. The best example of this is trout, which are adapted to high oxygen streams and cannot survive in water with DO content below the fishkill level: below 5 mg/L. This is an important point because when a stream ecosystem receives organic pollution it is especially susceptible to fishkill due to the corresponding reduction in oxygen levels.

Streams exhibit a large area for land-water interchange. Most streams are primarily detritus-based food chains. This means that their primary source of energy comes not from green plants, as in most ecosystems, but from organic material from the surrounding land, which is used as food by the decomposers. Smith (1974) points out that nutrients and waste products are transported by the flowing water to and away from many aquatic organisms; this process maintains a productivity that is many times greater than that in standing waters.

7.7 RUNNING WATERS AND THE HEAVY HAND OF MAN

Ecologically and geomorphologically, running water systems are dynamic, ever changing. On Earth, the biological functioning of rivers is naturally affected by changing geology and climate. These effects influence slope, basin, and channel characteristics of running water systems. Moreover, the character of stream bank vegetation is also affected. However, anthropogenic changes (man-made changes) now surpass natural changes both in rate and magnitude. The extent of the problem is well documented in the United States. For example, Benke (1990), in *A Perspective on America's Vanishing Streams,* reported that barely less than 2% of the 5+ million km of U.S. streams had sufficient high quality features to warrant regulatory protection. For those who have studied this problem, it is not surprising that by 2000 a large portion of the total streamflow of the world will be regulated. Rivers and their floodplains have been altered in many moderate to large rivers. A significant fraction of the aquatic biota of North America is imperiled. Without a doubt, running water systems are in need of preservation and, in some cases, restoration.

The majority of rivers in developed countries have been substantially affected by the construction of dams, channels, diversions, and other physical controls. It is important to note that when man alters natural stream flow, other things are also altered. As a case in point, consider the effects brought about by the installation of a dam in a large river system. The physical, chemical, and biological characteristics of the river below the dam differ greatly from free-flowing river before the dam.

What exactly does all this mean?

Well, for one thing, the natural variability in discharge and temperature are changed. In response to these changes, there is a shift in species composition in response to altered habitat. Even if the damming of a river brings about an increase in overall abundance of species, the richness of the species declines. To gain understanding of this problem, consider, for example, a classic problem with altering the natural course of streams and rivers: the plight of the salmon. Most people know that salmon are migratory fish. What most people do not know or do not think about is that when a dam is constructed in a river system, salmon and other migratory fish (e.g., steel-head) are sometimes barred from moving from the stream to the ocean and then back into stream to spawn. Obviously, this is not good for the salmon and is not good for those who depend on the salmon for their well-being, their livelihood. As the pressure grows to continue to divert running water flow, the regulators will need to specify minimum flow requirements and natural flow variation for healthy, functioning river systems (Allan, 1995).

It should be pointed out that the heavy hand of man, when he diverts stream flow, also alters the landscape through which the stream flows. When man spreads his settlements, and the settlements grow, the pressure put on running water systems is increased. For example, when a settlement along a river bank is initially made, the best land is taken first. As the settlement grows, the marginal land (left overs) such as floodplains, marshes, bogs, and other wetland sites are modified to suit man's needs. That is, the floodplains, marshes, and bogs are drained, the vital watershed is destroyed (trees cut down), agricultural projects spring up, and other less obvious projects are undertaken. The point is that what used to be a tranquil, quiet, natural running water setting, is now a development; compliments of the heavy hand of man.

Unfortunately, the story (tragedy) doesn't end here. Diverting stream flow also has an impact on important elements such as water tables, affecting both the level of water and the drainage aspects.

You know what? When you build or buy a house on a floodplain, you might get flooded out. Seems like a logical statement, right?

Have you checked out the flooding pattern of the river systems flowing through the upper Midwestern part of the United States lately? If you have, you probably get the message.

Another problem with diverting stream flow is erosion. When a stream course is altered, undisturbed landscape is disturbed, allowing more sediments to enter the stream. This increase in sediments has a detrimental effect on instream habitat. Removal of stream bank vegetation (riparian land) leads to a host of problems. For example, when vegetation is removed from a natural stream bank, the bank losses stability. The affected stream usually increases in temperature with a corresponding shift in the types of stream biota. Further, other activities of man, like dumping chemicals and organic waste into the stream, work to further exacerbate the situation.

It should be pointed out that when it comes to being contaminated, rivers are not defenseless. In fact, rivers, through their self-purification process, can (when given enough time and distance) purify themselves. However, it should also be pointed out that, as was stated in the opening of this chapter, rivers and streams do not have an infinite capacity for pollution.

7.8 BIOTIC INDICATORS OF POLLUTION

The diversity of species in an aquatic ecosystem is often a good indicator of the presence of pollution. The greater the diversity, the lower degree of pollution. The *biotic index* is used to determine the quality of a stream. The biotic index is a systematic survey of invertebrate aquatic organisms. It is based on two principles:

(1) Pollution tends to restrict the variety of organisms present at a point, although a large number of pollution-tolerant species may persist.

(2) In a polluted stream, as the degree of pollution increases, key organisms tend to disappear in the following order:
   • stone flies
   • mayflies
   • caddis flies
   • freshwater shrimp
   • tubificid worms

The biotic index ranges from 0 to 10. The most polluted aquatic environment, which therefore contains the smallest variety of organisms, is at the lowest end of the scale; the least polluted (cleanest) streams are at the highest.

Normally, in practice, streams have a wide range of biotic indices. As a rule of thumb a stream with a biotic index above 6 would support fish; below 4 it will not; and 1 or less is very toxic probably with only tubificid worms present. These tubificid worms (pollution worms) like to anchor themselves to the muck-filth stream bottom; just a writhing, wormy mass bending with the current.

CHAPTER 8

Water Quality

Are We To Wait Until All the Frogs "Croak"?

The earliest chorus of frogs—those high-pitched rhapsodies of spring peepers, those "jug-o-rum" calls of bullfrogs, those banjo-like bass harmonies of green frogs, those long and guttural cadences of leopard frogs, their singing a prelude to the splendid song of birds—beside an otherwise still pond on an early spring evening heralds one of nature's dramatic events: The drama of metamorphosis.

This metamorphosis begins with masses of eggs that soon hatch into gill-breathing, herbivorous, fishlike tadpole larvae. As they feed and grow, warmed by the spring sun, almost imperceptibly a remarkable transformation begins. Hindlegs appear and gradually lengthen. Tails shorten. Larval teeth vanish and lungs replace gills. Eyes develop lids. Forelegs emerge. In a matter of weeks the aquatic, vegetarian tadpole will (should it escape the many perils of the pond) complete its metamorphosis into an adult, carnivorous frog.

This springtime metamorphosis is special: this anticipated event (especially for the frog), marks the end of winter, the rebirth of life, a rekindling of hope (especially for man). This yearly miracle of change sums up in a few months each spring what occurred over 3,000 million years ago, when the frog evolved from its ancient predecessor. Today, however, something is different, strange, and wrong with this striking and miraculous event.

In the first place, where are all the frogs? Where have they gone? Why has their population decreased so dramatically in recent years?

The second problem: That this natural metamorphosis process (perhaps a reenactment of some paleozoic drama whereby, over countless generations, the first amphibian-types equipped themselves for life on land) now demonstrates aberrations of the worst kind, of monstrous proportions and dire results to frog populations in certain areas.

Consider, for example, information provided by Geraghty and Miller (1997) about the frog population that resides in the upper Midwestern section of the United States, specifically Minnesota. Reports have surfaced with accounts of deformed frogs that have been found at more than 100 sites in 54 of 87 Minnesota counties. Moreover, USEPA has received many similar reports from the U.S. and Canada as well as parts of Europe.

Most deformities have been in rear legs and appear to be developmental. The question is: Why?

Researchers have noted that neurological abnormalities have also been found. Again, the question is why?

Researchers have pointed the finger of blame at parasites, pesticides, and other chemicals, ultraviolet radiation, acid rain, and metals. Something is going on. What is it? We do not know!

The next question becomes: What are we going to do about it? Are we to wait until all the frogs croak before we act—before we find the source, the cause, the polluter—before we see this reaction result in other species; maybe in our own?

The final questions are obvious: When frogs are forced by mutation into something else, is this evolution by gunpoint?

Is man holding the gun?

8.1 INTRODUCTION

BECAUSE the amount of rain and snow remains almost constant, but population and usage per person are both increasing rapidly, water is in short supply. In the United States alone, water usage is four times greater today than it was in 1900. In the home, this increased use is directly related to an increase in the number of bathrooms, garbage disposals, home laundries, and lawn sprinklers. In industry, usage has increased 13 times since 1900.

Drinking water is provided to approximately 200+ million Americans by 60,000+ community water supply systems, and to nonresidential locations, such as schools, factories, campgrounds, by 170,000 small scale suppliers. The rest of Americans are served by private wells. About 90% of the drinking water used in the U.S. is supplied from groundwater. Untreated water drawn from groundwater and surface waters, and used as a drinking water supply, can contain contaminants that pose a threat to human health.

Obviously, with a limited amount of drinking water available for use, water that is available must be reused or we will be faced with an inadequate supply to meet the needs of all users. Water reuse is complicated by water pollution. Pollution is relative and hard to define. For example, floods and dead animals are polluters, but their effects are local and tend to be temporary. Today, water is polluted in many sources, and pollution exists in many forms. It may appear as excess aquatic weeds, oil slicks, a decline in sport fishing, and an increase in carp, sludge worms, and other forms of life that readily tolerate pollution. Maintaining water quality is important because water pollution is not only detrimental to health but also to recreation, commercial fishing, aesthetics, and private, industrial, and municipal water supplies.

At this point in this discussion you might be asking yourself a question: With all the recent publicity about pollution and the enactment of new environmental regulations, hasn't water quality in the United States improved recently?

Well, with the recent pace of achieving fishable-swimmable waters under the Clean Water Act (CWA), you might think so.

In 1994 the *National Water Quality Inventory Report to Congress* indicated that 63% of the nations' lakes, rivers, and estuaries meet designated uses—only a slight increase over that reported in 1992.

The culprit is primarily *nonpoint source pollution* (NPS) (to be discussed later). NPS is the leading cause of impairment for rivers, lakes, and estuaries. Impaired sources are those that do not fully support designated uses, such as fish consumption, drinking water supply, groundwater recharge, aquatic life support, or recreation. According to Fortner & Schechter (1996), the five leading sources of water quality impairment in rivers are: agriculture, municipal wastewater treatment plants, habitat and hydrologic modification, resource extraction, and urban runoff and storm sewers.

Thus, in attempting to answer the original question—hasn't water quality in the United States improved recently?—the best answer probably is: We are holding our own in controlling water pollution, but we need to make more progress. This understates an important point; that is, when it comes to water quality, we need to make more progress on a continuing basis.

8.2 WATER QUALITY: REGULATORY REQUIREMENTS

The effort to regulate drinking water and wastewater effluent has increased since the early 1900s. Beginning with an effort to control the discharge of wastewater into the environment, preliminary regulatory efforts focused on protecting public health. The goal of this early wastewater treatment program was to remove suspended and floatable material, treat biodegradable organics, and eliminate pathogenic organisms. Thus, regulatory efforts were pointed toward constructing wastewater treatment plants in an effort to alleviate the problem. But then a problem soon developed: Progress. Progress in the sense that time marched on and with it so did cities in the United States where it became increasingly difficult to find land required for wastewater treatment and disposal. Wastewater professionals soon recognized the need to develop methods of treatment that would accelerate "nature's way" (the natural purification of water) under controlled conditions in treatment facilities of comparatively smaller size.

Regulatory influence on water-quality improvements in both drinking water and wastewater took a giant step forward in the 1970s. The Safe Drinking Water Act (SDWA), passed by Congress in 1974, started a new era in the field of drinking water supply to the public. At about the same time, wastewater treatment, under the Water Pollution Control Act Amendments of 1972, (Clean Water Act, CWA), established national water pollution control goals.

8.2.1 SAFE DRINKING WATER ACT OF 1974

The Safe Drinking Water Act of 1974 mandated the USEPA to establish drinking-water standards for all public water systems serving 25 or more people or having 15 or more connections. Pursuant to this mandate, EPA has established maximum contaminant levels for drinking water delivered through public water distribution systems. The maximum contaminant levels (MCL) of inorganics, organic chemicals, turbidity, and microbiological contaminants are shown in Table 8.1. The EPA's primary regulations are mandatory and must be complied with by all public water systems to which they apply. If analysis of the water produced by a water system indicates that an MCL for a contaminant is being exceeded, the system must take steps to stop providing the water to the public or initiate treatment to reduce the contaminant concentration to below the MCL.

EPA has also issued guidelines to the states with regard to secondary drinking-water standards. These appear in Table 8.2. These guidelines apply to drinking water contaminants that may adversely affect the aesthetic qualities of water, such as odor and appearance. These qualities have no known adverse health effects, and thus secondary regulations are not mandatory. However, most drinking-water systems comply with the limits; they have learned through experience that the odor and appearance of drinking water is not a problem until customers complain. One thing is certain, they will complain and quite often.

The question is: Can you blame them (us)?

8.2.2 WATER POLLUTION CONTROL ACT OF 1972
(CLEAN WATER ACT, CWA)

The Water Pollution Control Act of 1972 mandated the EPA to establish standards for wastewater discharges. Before enactment of this public law, (PL 92-500), there were no *specific* national water pollution control goals

TABLE 8.1. EPA Primary Drinking Water Standards.

1. Inorganic Contaminant Levels	
Contaminate	Level (mg/L)
Arsenic	0.05
Barium	1.00
Cadmium	0.010
Chromium	0.05
Lead	0.05
Mercury	0.002
Nitrate	10.00
Selenium	0.01
Silver	0.05

2. Organic Contaminant Levels	
Chemical	Maximum Contaminant Level (MCL) mg/L
Chlorinated hydrocarbons:	
Endrin	0.0002
Lindane	0.004
Mexthoxychlor	0.1
Toxaphene	0.005
Chlorophenoxys:	
2,4-D	0.1
2,4,5-TP silvex	0.01

3. Maximum Levels of Turbidity	
Reading Basis	Maximum Contaminant Level (MCL) turbidity units
Turbidity reading (monthly average)	1 TU or up to 5 TUs if the water supplier can demonstrate to the state that the higher turbidity does not interfere with disinfection, maintenance of an effective disinfectant agent throughout the distribution system, or microbiological determinants.
Turbidity reading (based on average of two consecutive days)	5 TUs

TABLE 8.1 (continued). EPA Primary Drinking Water Standards.

4. Microbiological Contaminants			
		Individual Sample Basis	
Test Method Use	Monthly Basis	Fewer than 20 Samples/mo	More than 20 Samples/mo
Membrane filter technique	1/100 mL average daily	Number of coliform bacteria not exceed:	
		4/100 mL in more than 1 sample	4/100 mL in more than 5% of samples
Fermentation		Coliform bacteria shall not be present in:	
10-mL standard portions	More than 10% of the portions	3 or more portions in more than 1 sample	3 or more portions in more than 5% of samples
100-mL standard portions	More than 60% of the portions	5 portions in more than 1 sample	5 portins in more than 20% of the samples

Source: Adapted from USEPA: National Interim Primary Drinking Water Regulations, *Federal Register,* part IV, December 24, 1975.

TABLE 8.2. Secondary Maximum Contaminant Levels.

Contaminant	Level	Adverse Effect
Chloride	250 mg/L	Causes taste
Color	15 cu	Appearance problems
Copper	1 mg/L	Metallic taste
Corrosivity	noncorrosive	Tastes and odors
Fluoride	2 mg/L	Dental fluorosis
Foaming agents	0.5 mg/L	Appearance problems
Iron	0.3 mg/L	Appearance problems
Manganese	0.05 mg/L	Discolors laundry
Odor	3 TON	Unappealing to drink
pH	6.5–8.5	Corrosion or scaling
Sulfate	250 mg/L	Has laxative effect
Total dissolved solids	500 mg/L	Taste, corrosion
Zinc	5 mg/L	Taste, appearance

cu = color unit.
TON = threshold odor number.
Source: Adapted from McGhee, (1991), p. 161.

162

or objectives. Current standards require that municipal wastewater be given secondary treatment (to be explained in detail later) and that most effluents meet the conditions shown in Table 8.3. The goal, via secondary treatment, was set in order that the principal components of municipal wastewater, suspended solids, biodegradable material, and pathogens could be reduced to acceptable levels. Industrial dischargers are required to treat their wastewater to the level obtainable by the *best available technology* (BAT) for wastewater treatment in that particular type of industry.

Moreover, a National Pollution Discharge Elimination System (NPDES) program was established based on uniform technological minimums with which each point source discharger has to comply (Metcalf & Eddy, 1991). Under NPDES, each municipality and industry discharging effluent into streams is assigned discharge permits. These permits reflect the secondary treatment and best-available-technology standards.

Before beginning a discussion of the characteristics of water and wastewater and the factors affecting water quality, this is a good place to answer a pertinent question: Just exactly what is water quality?

In this text *water quality* refers to those characteristics or range of characteristics that make water appealing and useful. Keep in mind that *useful* also means nonharmful or nondisruptive to either ecology or the human condition within the very broad spectrum of possible uses of water. For example, the absence of odor, turbidity, or color are desirable immediate qualities. However, there are imperceptible qualities that are also important; that is, the chemical qualities. The fact is the presence of materials such as toxic metals, (e.g., mercury and lead), excessive nitrogen and phosphorous, or dissolved organic material may not be readily perceived by the senses, but may exert substantial negative impacts on the health of a stream and/or on human health. The ultimate impact of these imperceptible qualities of water (chemicals) on the user may be nothing more than loss of aesthetic values. On the other hand, water containing chemicals could also lead to a reduction in biological health or to an outright degradation of human health.

Simply stated: In the science of water, the importance of water quality cannot be overstated.

TABLE 8.3. Minimum National Standards for Secondary Treatment.

Characteristic of Discharge	Unit of Measure	Average 30-day Concentration	Average 7-day Concentration
BOD$_5$	mg/L	30	45
Suspended solids	mg/L	30	45
Hydrogen-ion concentration	pH units	Within the range of 6.0 to 9.0	

Source: Federal Register, "Secondary Treatment Regulations," 40 CFR Part 133, July 1, 1988.

8.3 CHARACTERISTICS OF WATER AND WASTEWATER

A knowledge of the parameters/characteristics most commonly associated with water and wastewater treatment processes is essential to the water/wastewater specialist. Therefore, the remainder of this chapter will be devoted to a discussion of parameters used to assess the physical, chemical, and biological characteristics of water and wastewater.

Before beginning a discussion of the physical water quality parameters/characteristics it should be pointed out that when this text refers to water quality, the definition used is predicated on the intended use of the water. The point is that many parameters have evolved that qualitatively reflect the impact that various contaminants (impurities) have on selected water uses; the following sections provide a brief discussion of these parameters.

8.3.1 PHYSICAL CHARACTERISTICS OF WATER/WASTEWATER

Probably the most apparent physical characteristic of water is that it is wet. Have you ever wondered what makes water wet? David Clary (1997), a chemist at University College London, points out that water does not start to behave like a liquid until at least six molecules form a cluster. He found that groups of five water molecules or fewer have planar structures, forming films one molecule thick. However, when a sixth molecule is added, the cluster switches to a three-dimensional cage-like structure and suddenly it has the properties of water; that is, it is all wet.

The information provided above is all well and good (we all know that water is wet), but the other physical characteristics of water/wastewater that we are interested in here are somewhat more germane to the discussion at hand; namely, a category of parameters/characteristics that can be used to define water quality. One such category is the physical characteristics for water that respond to the senses of smell, taste, sight, and touch. Solids, turbidity, color, taste and odor, and temperature also fall into this category.

8.3.1.1 Solids

Other than gases, all contaminants of water contribute to the solids content. Classified by their size and state, by their chemical characteristics, and by their size distribution, solids can be dispersed in water in both suspended and dissolved forms. In regards to size, solids in water and wastewater can be classified as suspended, settleable, colloidal, or dissolved. Solids are also characterized as being *volatile* or *nonvolatile*. The distribution of solids is determined by computing the percentage of filtrable solids by size range. Solids typically include inorganic solids such as silt and clay from riverbanks and organic matter such as plant fibers and microorganisms from natural or man-made sources. In flowing water, many of these contaminants result from the erosive action of water flowing over surfaces. It should be pointed out that suspended material is not normally found in groundwater. This is the case because of the filtering effect imparted by the soil.

In water, suspended material is objectionable because it provides adsorption sites for biological and chemical agents. These adsorption sites provide attached microorganisms a protective barrier against the chemical action of chlorine. In addition, suspended solids in water may be degraded biologically resulting in objectionable by-products. Thus, the removal of these solids is of great concern in the production of clean, safe drinking water and wastewater effluent.

In water treatment, the most effective means of removing solids from water is by filtration. It should be pointed out, however, that not all solids, such as colloids and other dissolved solids, can be removed by filtration.

In wastewater treatment, suspended solids is an important water-quality parameter and is used to measure the quality of the wastewater influent, to monitor performance of several processes, and to measure the quality of effluent. As shown in Table 8.3, USEPA has set a maximum suspended-solids standard of 30 mg/L for most treated wastewater discharges.

8.3.1.2 Turbidity

One of the first things that is noticed about water is its clarity. The clarity of water is usually measured by its *turbidity*. Turbidity is a measure of the extent to which light is either absorbed or scattered by suspended material in water. Absorption and scattering are influenced by both the size and surface characteristics of the suspended material.

In surface water, most turbidity results from the erosion of very small colloidal material such as rock fragments, silt, clay, and metal oxides from the soil. Microorganisms and vegetable material may also contribute to turbidity. Wastewaters from industry and households usually contain a wide variety of turbidity-producing materials. Detergents, soaps, and various emulsifying agents contribute to turbidity.

In water treatment, turbidity is useful in defining drinking-water quality. In wastewater treatment, turbidity measurements are particularly important whenever ultraviolet radiation (UV) is used in the disinfection process. For UV light to be effective in disinfecting wastewater effluent, UV light must be able to penetrate the stream flow. Obviously, stream flow that is turbid works to reduce the effectiveness of irradiation (penetration of light).

The colloidal material associated with turbidity provides adsorption sites for microorganisms and chemicals that may be harmful or cause undesirable tastes and odors. Moreover, the adsorptive characteristics of many colloids work to provide protection sites for microorganisms from disinfection processes. Turbidity in running waters interferes with light penetration and photosynthetic reactions.

8.3.1.3 Color

Color is another physical characteristic by which the quality of water can be judged. Pure water is colorless. Water takes on color when foreign substances such as organic matter from soils, vegetation, minerals, and aquatic organisms are present. Color can also be contributed to water by municipal and industrial wastes.

Color in water is classified as either *true color* or *apparent color*. Water whose color is partly due to dissolved solids that remain after removal of suspended matter is known as true color. Color contributed by suspended matter is said to have apparent color. In water treatment, true color is the most difficult to remove.

The obvious problem with colored water is that it is not acceptable to the general public. That is, given a choice, the public prefers clear, uncolored water. Another problem with colored water is the affect it has on laundering, papermaking, manufacturing, textiles, and food processing. The point is that the color of water has a profound effect on its marketability for both domestic and industrial use.

In water treatment, color is not usually considered unsafe or unsanitary, but is a treatment problem in regards to exerting a chlorine demand, which reduces the effectiveness of chlorine as a disinfectant.

In wastewater treatment, color is not necessarily a problem, but instead is an indicator of the *condition* of the wastewater. Condition refers to the age of the wastewater, which, along with odor, provides a qualitative indication of its age. Early in the flow, wastewater is a light brownish-gray color. As the travel time in the collection system increases (flow becomes increasingly more septic), and more anaerobic conditions develop, the color of the wastewater changes from gray to dark gray and ultimately to black.

8.3.1.4 Taste and Odor

Taste and odor are used jointly in the vernacular of water science. Odor is used in wastewater, taste is not a consideration, thank you very much! As stated previously, in drinking water, taste and odor are not normally a problem until the consumer complains. The problem is, of course, that most consumers find taste and odor in water aesthetically displeasing. Taste and odor do not directly present a health hazard, but they can cause the customer to seek water that tastes and smells good, but may not be safe to drink. The fact is most consumers consider water to be tasteless and odorless. Thus, when the consumer finds that his/her drinking water has a taste or odor, or both, he/she automatically associates the drinking water with contamination.

Water "contaminants" are attributable to contact with nature or human use. Taste and odor in water are caused by a variety of substances such as minerals, metals, and salts from the soil, constituents of wastewater, and end products produced in biological reactions. When water has a taste but no accompanying odor, the cause is usually inorganic contamination. Water that tastes bitter is usually alkaline, while salty water is commonly the result of metallic salts. However, when water has both taste and odor, the likely cause is organic materials. The list of possible organic contaminants is too long to record here, however petroleum-based products lead the list of offenders. Taste- and odor-producing liquids and gases in water are produced by biological decomposition of organics. A prime example of one of these is hydrogen sulfide, known best for its characteristic "rotten-egg" taste and odor. Certain species of algae also secrete an oily substance that may produce both taste and odor. When certain substances combine (such as organics and chlorine), the synergistic effect produces taste and odor.

In wastewater, odors are of major concern, especially to those who reside in close proximity to a wastewater treatment plant. These odors are generated by gases produced by decomposition of organic matter or by substances added to the wastewater. Since these substances are fairly volatile, they are readily released to the atmosphere at any point where the waste stream is exposed, particularly if there is turbulence at the surface.

Most people would argue that all wastewater is the same; it has a disagreeable odor. It is hard to argue against the disagreeable odor. However, one wastewater operator told me that wastewater "smelled great—smells just like money to me—money in the bank," she said. It should be pointed out, however, that there is a difference in odor between different types of wastewater. For example, fresh wastewater has a distinctive odor which is less objectionable than the odor of wastewater that has undergone anaerobic decomposition.

In water treatment, one of the common methods used to remove taste and odor is to oxidize the materials that cause the problem. Oxidants, such as potassium permanganate and chlorine, are used. Another common treatment method is to feed powdered activated carbon prior to the filter. The activated carbon has numerous small openings that adsorb the components that cause the odor and tastes.

In wastewater treatment, odor control is a never-ending problem. To combat this difficult problem, odors must be contained. In most urban plants it has become necessary to physically cover all source areas such as treatment basins, clarifiers, aeration basins, and contact tanks to prevent odors from leaving the processes. These contained spaces must then be positively vented to wet-chemical scrubbers to prevent the buildup of toxic concentrations of gas.

8.3.1.5 Temperature

Heat is added to surface and groundwater in many ways. Some of these are natural, some artificial. The problem with heat or temperature in surface waters, for example, is that it affects the solubility of oxygen in water, the rate of bacterial activity, and the rate at which gases are transferred to and from the water.

It is important to point out that in the examination of water or wastewater, temperature is not normally used to evaluate either. However, temperature is one of the most important parameters in natural surface-water systems. Surface waters are subject to great temperature variations.

Water temperature determines, in part, how efficiently certain water treatment processes operate. For example, temperature has an effect on the rate at which chemicals dissolve and react. When water is cold, more chemicals are required for efficient coagulation and flocculation to take place. When water temperature is high, the result may be a higher chlorine demand because of the increased reactivity and also because there is often an increased level of algae and other organic matter in raw water. Temperature also has a pronounced effect on the solubilities of gases in water.

Ambient temperature (temperature of the surrounding atmosphere) has the most profound and universal effect on temperature of shallow natural water systems. When water is used by industry to dissipate process waste heat, the discharge points into surface waters may experience localized temperature changes that are quite dramatic. Other sources of increased temperatures in running water systems result because of clear-cutting practices in forests (where protective canopies are removed) and also from irrigation flows returned to a body of running water.

In wastewater treatment, the temperature of wastewater is generally warmer than that of the water supply. This is the case because of the addition of warm water from industrial activities and households. In the treatment process itself, temperature not only influences the metabolic activities of the microbial population but also has a profound effect on such factors as gas-transfer rates and the settling characteristics of the biological solids (Metcalf & Eddy, 1991).

8.3.2 CHEMICAL CHARACTERISTICS OF WATER

Another category of parameters that can be used to define water quality is its chemical characteristics. The principal chemical constituents found in

TABLE 8.4. Chemical Constituents Commonly Found in Water.

Constituent	
Calcium	Fluorine
Magnesium	Nitrate
Sodium	Silica
Potassium	TDS
Iron	Hardness
Manganese	Color
Bicarbonate	pH
Carbonate	Turbidity
Sulfate	Temperature
Chloride	

water are shown in Table 8.4. These chemical constituents are important because each one affects water use in some manner; each one either restricts or enhances specific uses.

Let's take a look at one example. The pH of water is important. As pH rises, for example, the equilibrium (between bicarbonate and carbonate) increasingly favors the formation of carbonate, which often results in the precipitation of carbonate salts. If you have ever had flow in a pipe system interrupted or a heat-transfer problem in your water heater system, then you most likely have been victimized by carbonate salts that formed a hard-to-dissolve scale within the system. It should be pointed out that not all carbonate salts have a negative effect on their surroundings. Consider, for example, the case of those blue marl lakes; they owe their unusually clear appearance to carbonate salts.

Earlier it was pointed out that water has been called the *universal solvent*. This is, of course, a fitting description. The solvent capabilities of water are directly related to its chemical characteristics or parameters.

In water-quality management, total dissolved solids (TDS), alkalinity, hardness, fluorides, metals, organics, and nutrients are the major chemical parameters of concern.

8.3.2.1 Total Dissolved Solids

Total Dissolved Solids (TDS) is a function of the minerals dissolved from rocks and soil as water passes over and through it. TDS constitutes a part of total solids in water; it is the material remaining in water after filtration.

Dissolved solids may be organic or inorganic. Water may come into contact with these substances within the soil, on surfaces, and in the atmosphere. The organic dissolved constituents of water come from the decay products of vegetation, from organic chemicals, and from organic gases.

Dissolved solids can be removed from water by distillation, electrodialysis, reverse osmosis, or ion exchange. It is desirable to remove these dissolved minerals, gases, and organic constituents because they may cause physiological effects and produce aesthetically displeasing color, taste, and odors.

While it is desirable to remove many of these dissolved substances from water, it is not prudent to remove them all. This is the case, for example, because pure, distilled water has a flat taste. Further, water has an equilibrium state with respect to dissolved constituents. Thus, if water is out of equilibrium or undersaturated, it will aggressively dissolve materials it comes into contact with. Because of this problem, substances that are readily dissolvable are sometimes added to pure water to reduce its tendency to dissolve plumbing.

8.3.2.2 Alkalinity

Another important characteristic of water is its *alkalinity*, which is a measure of the water's ability to absorb hydrogen ions without significant pH change. Simply stated, alkalinity is a measure of the buffering capacity of water. Alkalinity is thus a measure of the ability of water to neutralize acids. The major chemical constituents of alkalinity in natural water supplies are the bicarbonate, carbonate, and hydroxyl ions. These compounds are mostly the carbonates and bicarbonates of sodium, potassium, magnesium, and calcium. These constituents originate from carbon dioxide (from the atmosphere and as a by-product of microbial decomposition of organic material) and from their mineral origin (primarily from chemical compounds dissolved from rocks and soil).

Highly alkaline waters are unpalatable; however, this condition has little known significance for human health. The principal problem with alkaline water is the reactions that occur between alkalinity and certain substances in the water. The resultant precipitate can foul water system appurtenances. Also, alkalinity levels affect the efficiency of certain water treatment processes, especially the coagulation process.

8.3.2.3 Hardness

If you have ever vigorously washed your hands with a bar of soap and noticed that you needed more soap to "get a lather," you have encountered another important chemical characteristic of water: *hardness*. In its simplest terms, hardness is the presence in water of multivalent cations, most notably calcium and magnesium ions. Hardness is classified as *carbonate hardness* and *noncarbonate hardness.* The carbonate that is equivalent to the alkalinity is termed carbonate hardness. Hardness is either temporary or permanent. Carbonate hardness (temporary hardness) can be removed by boiling. Noncarbonate hardness cannot be removed by boiling and is classified as permanent.

Hardness values are expressed as an equivalent amount or equivalent weight of calcium carbonate (*equivalent weight* of a substance is its atomic or molecular weight divided by *n*). Water with a hardness of less than 50 ppm is soft. Above 200 ppm, domestic supplies are usually blended to reduce the hardness value. The U.S. Geological Survey uses the following classification:

Range of Hardness [mg/liter (ppm) as $CaCO_3$]	Descriptive Classification
1 to 50	Soft
51 to 150	Moderately hard
151 to 300	Hard
Above 300	Very hard

The impact of hardness can be measured in economic terms. The example about soap consumption points this out; that is, soap consumption represents an economic loss to the water user. There is another problem with soap and hardness. When one uses a bar of soap in hard water, he/she works the soap until a lather is built up. When lathering does occur, the water has been "softened" by the soap. The problem is that the precipitate formed by hardness and soap (soap curd) adheres to just about anything (tubs, sinks, dishwashers) and may stain clothing, dishes, and other items. There also is a personal problem: the residues of the hardness-soap precipitate may remain in the pores, causing skin to feel rough and uncomfortable. Today these problems have been largely reduced by the development of synthetic soaps and detergents that do not react with hardness. However, hardness still leads to other problems, scaling and laxative effect. As pointed out earlier, scaling occurs when carbonate hard water is heated and calcium carbonate and magnesium hydroxide are precipitated out of solution, forming a rock-hard scale that clogs hot-water pipes and reduces the efficiency of boilers, water heaters, and heat exchangers. Hardness, especially with the presence of magnesium sulfates, can lead to the development of a laxative effect on new consumers.

There are advantages to be gained from usage of hard water. These include: (1) hard water aids in growth of teeth and bones; (2) hard water reduces toxicity to man by poisoning with lead oxide from pipelines made of lead; and (3) soft waters are suspected to be associated with cardiovascular diseases (Rowe & Abdel-Magid, 1995).

8.3.2.4 Fluoride

How many fifty-something parents have looked at the quality of the teeth of their thirty-something offspring with amazement and envy?

This might seem like a strange statement to some readers, but consider this: these fifty-something parents of today grew up in the 1940s and 1950s. These were the years when their teeth were growing, maturing. During these formative years, very few members of this age group were exposed to fluoride in drinking water. There are, of course, those who belong to the fifty-something group who have excellent teeth. These individuals practiced good dental hygiene, lived in areas where hard water was available, and they may have lived in a geographic area where a small amount of natural fluoride was in their drinking water. The point is that with the addition of fluoride to most local drinking water supplies in the late 1950s, many people who grew up during this period have very good teeth—thanks to fluoride.

Fluoride is seldom found in appreciable quantities in surface waters and appears in groundwater in only a few geographical regions. However, fluoride is sometimes found in a few types of igneous or sedimentary rocks. Fluoride is toxic to humans in large quantities. Fluoride is also toxic to some animals. For example, certain plants used for fodder have the ability to store and concentrate fluoride. When animals consume this forage, they ingest an enormous overdose of fluoride. Animals' teeth become mottled, they lose weight, give less milk, grow spurs on their bones, and become so crippled they must be destroyed (Koren, 1991).

As pointed out earlier, fluoride, used in small concentrations (about 1.0 mg/L in drinking water), can be beneficial. Experience has shown that drinking water containing a proper amount of fluoride can reduce tooth decay by 65% in children between ages 12 and 15.

How does fluoridization of a drinking water supply actually work to reduce tooth decay?

Fluoride combines chemically with tooth enamel when permanent teeth are forming. The result, of course, is teeth that are harder, stronger, and more resistant to decay.

When large concentrations are used (>2.0 mg/L), discoloration of teeth may result. Adult teeth are not affected by fluoride. EPA sets the upper limits for fluoride based on ambient temperatures. This is the case because people drink more water in warmer climates; therefore, fluoride concentrations should be lower in these areas.

8.3.2.5 Metals

Although iron and manganese are most commonly found in groundwaters, surface waters may also contain significant amounts at times. Metals in water are classified as either nontoxic or toxic. Only those metals that are harmful in relatively small amounts are labeled toxic; other metals fall into the nontoxic group. In natural waters, sources of metals include dissolution from natural deposits and discharges of domestic, agricultural, or industrial wastewaters.

Nontoxic metals commonly found in water include the hardness ions, calcium and magnesium, iron, manganese, aluminum, copper, zinc, and sodium. Sodium, which is abundant in the earth's crust and is highly reactive with other elements, is by far the most common nontoxic metal found in natural waters. The salts of sodium, in excessive concentrations, cause a bitter taste in water and are a health hazard to kidney and cardiac patients. Sodium, in large concentrations, is toxic to plants.

Although iron and manganese in natural waters, in very small quantities, may cause color problems, they frequently occur together and present no health hazard at normal concentrations. There are some bacteria, however, that use iron and manganese compounds for an energy source, and the resulting slime growth may produce taste and odor problems.

Iron exists in two forms: ferrous and ferric. Ferrous iron is found in well waters, groundwaters or bottom layers of stratified lakes, and in other waters with a low level of dissolved oxygen. Under anaerobic conditions, waters can have significant dissolved-iron concentrations.

Manganese creates problems in a water system similar to those created by iron. In surface waters, manganese concentrations seldom reach high levels. However (like iron), in groundwaters subject to anaerobic conditions, manganese concentrations can become very high.

In natural water systems, other nontoxic metals are generally found in very small quantities. Most of these metals cause taste problems well before they reach toxic levels.

Fortunately, toxic metals are present in only minute quantities in most natural water systems. However, even in small quantities, toxic metals in drinking water are harmful to humans and other organisms. Arsenic, barium, cadmium, chromium, lead, mercury, and silver are toxic metals that may be dissolved in water. Arsenic, cadmium, lead, and mercury, all cumulative toxins, are particularly hazardous. These particular metals are concentrated by the food chain, thereby posing the greatest danger to organisms near the top of the chain.

8.3.2.6 Organics

According to Tchobanoglous and Schroeder (1987), the presence of organic matter in water is troublesome for the following reasons: "(1) color formation, (2) taste and odor problems, (3) oxygen depletion in streams, (4) interference with water treatment processes, and (5) the formation of halogenated compounds when chlorine is added to disinfect water" (p. 94).

Generally, the source of organic matter in water is from decaying leaves, weeds, and trees; the amount of these materials present in natural waters is usually low. The general category of "organics" in natural waters includes organic matter whose origins could be from both natural sources and from human activities. It is important to distinguish natural organic compounds from organic compounds that are solely man-made (anthropogenic), such as pesticides and other synthetic organic compounds.

Many organic compounds are soluble in water, and surface waters are more prone to contamination by natural organic compounds than are groundwaters. In water, dissolved organics are usually divided into two categories: *biodegradable* and *nonbiodegradable.*

Biodegradable (break down) material consists of organics that can be utilized for nutrients (food) by naturally occurring microorganisms within a reasonable length of time. These materials usually consist of alcohols, acids, starches, fats, proteins, esters, and aldehydes. They may result from domestic or industrial wastewater discharges, or they may be end products of the initial microbial decomposition of plant or animal tissue. The principal problem associated with biodegradable organics is the effect resulting from the action of microorganisms. Moreover, some biodegradable organics can cause color, taste, and odor problems.

Oxidation and *reduction* play an important accompanying role in microbial utilization of dissolved organics. In oxidation, oxygen is added, or hydrogen is deleted from elements of the organic molecule. Reduction occurs when hydrogen is added to or oxygen is deleted from elements of the organic molecule. The oxidation process is by far more efficient and is predominant when oxygen is available. In *oxygen-present (aerobic)* environments, the end products of microbial decomposition of organics are stable and acceptable compounds. On the other hand, *oxygen-absent (anaerobic)* decomposition results in unstable and objectionable end products.

The quantity of oxygen-consuming organics in water is usually determined by measuring the *biological oxygen demand (BOD):* the amount of dissolved oxygen needed by aerobic decomposers to break down the organic materials in a given volume of water over a 5-day incubation period at 20°C (68°F).

Nonbiodegradable organics are resistant to biological degradation. For example, constituents of woody plants such as tannin and lignic acids, phenols, and cellulose are found in natural water systems and are considered refractory (resistant to biodegradation). In addition, some polysaccharides with exceptionally strong bonds and benzene with its ringed structure are essentially nonbiodegradable. An example is benzene associated with the refining of petroleum.

Some organics are toxic to organisms and thus are nonbiodegradable. These include the organic pesticides and compounds that have combined with chlorine.

Pesticides and herbicides have found widespread use in agriculture, forestry (silviculture), and mosquito control. Surface streams are contaminated via runoff and washoff by rainfall. These toxic substances are harmful to some fish, shellfish, predatory birds, and mammals. Some compounds are toxic to humans.

Certain nonbiodegradable chemicals can react with oxygen dissolved in water. The *chemical oxygen demand (COD)* is a more complete and accurate measurement of the total depletion of dissolved oxygen in water.

8.3.2.7 Nutrients (Biostimulents)

Nutrients are elements, such as carbon, nitrogen, phosphorous, sulfur, calcium, iron, potassium, manganese, cobalt, and boron, that are essential to the growth and reproduction of plants and animals. On the one hand, aquatic species depend on the surrounding water to provide their nutrients. On the other hand, nutrients, in terms of water quality, can be considered as pollutants when their concentrations are sufficient to encourage excessive growth of aquatic plants (e.g., algal blooms). As pointed out above, a wide variety of elements can be classified as nutrients, however, those required in most abundance by aquatic species are carbon, nitrogen, and phosphorous. Plants require large amounts of each of these three nutrients, otherwise growth will be *limited*.

Carbon is readily available from a number of natural sources including alkalinity, decaying products of organic matter, and from dissolved carbon dioxide from the atmosphere. Since carbon is readily available, it is seldom the *limiting nutrient.* This is an important point because it suggests that algal growth can be controlled by identifying and reducing the supply of a particular nutrient. In most cases, nitrogen and phosphorous are essential growth factors and are the limiting factors in aquatic plant growth. According to Welch (1980), seawater is most often limited by nitrogen, while freshwater systems are most often limited by phosphorus.

Nitrogen gas (N_2), which is extremely stable, is the primary component of the earth's atmosphere. Major sources of nitrogen include runoff from animal feedlots, fertilizer runoff from agricultural fields, from municipal wastewater discharges, and from certain bacteria and blue-green algae that can obtain nitrogen directly from the atmosphere. In addition, certain forms of acid rain can also contribute nitrogen to surface waters.

Nitrogen in water is commonly found in the form of nitrate (NO_3). Nitrate in drinking water can lead to a serious problem. Specifically, nitrate poisoning in infant humans, including animals, can cause serious problems and even death. This is the case because of a bacteria commonly found in the intestinal tract of infants that can convert nitrate to highly toxic nitrites (NO_2). Nitrite can replace oxygen in the bloodstream and result in oxygen starvation that causes a bluish discoloration of the infant ("blue baby" syndrome).

In aquatic environments, phosphorous is found in the form of phosphate. Major sources of phosphorous include phosphates in detergents, fertilizer and feedlot runoff, and municipal wastewater discharges.

8.3.3 CHEMICAL CHARACTERISTICS OF WASTEWATER

The following discussion of the chemical characteristics of wastewater is presented in three parts: (1) organic matter, (2) inorganic matter, and (3) gases. Metcalf & Eddy (1991) point out that in "wastewater of medium strength, about 75% of the suspended solids and 40% of the filterable solids are organic in nature" (p. 65). The organic substances of interest in this discussion include proteins, oil and grease, carbohydrates, and detergents (surfactants).

8.3.3.1 Organic Substances

Proteins are nitrogenous organic substances of high molecular weight found in the animal kingdom and to a lesser extent in the plant kingdom. The amount present varies from a small percentage found in tomatoes and other watery fruits and in the fatty tissues of meat, to quite a high percentage in lean meats and beans. The point is that all raw foodstuffs, plant and animal, contain proteins. Proteins consist wholly or partially of very large numbers of amino acids. They also contain carbon, hydrogen, oxygen, sulfur, phosphorous, and a fairly high and constant proportion of nitrogen. The molecular weight of proteins is quite high.

Coackley (1975) points out that proteinaceous materials constitute a large part of the wastewater biosolids, and that the biosolids particles, if they do not consist of pure protein, will be covered with a layer of protein that will govern their chemical and physical behavior. Moreover, the protein content ranges between 15 to 30% of the organic matter present for digested biosolids, and 28 to 50% in the case of activated biosolids. Proteins and urea are the chief sources of nitrogen in wastewater. When proteins are present in large quantities, microorganisms decompose them producing end products that have objectionable foul odors. During this decomposition process, proteins are hydrolyzed to amino acids, then further degraded to ammonia, hydrogen sulfide, and to simple organic compounds.

Oils and *grease* are another major component of foodstuffs. They are insoluble in water but dissolve in organic solvents such as petroleum, chloroform, and ether. Fats, oils, waxes, and other related constituents found in wastewater are commonly grouped under the term "grease." Fats and oils are contributed to domestic wastewater in butter, lard, margarine, and vegetable fats and oils (Metcalf & Eddy, 1991). Fats, which are compounds of alcohol and glycerol, are among the more stable of organic compounds and are not easily decomposed by bacteria. However, they can be broken down by mineral acids resulting in the formation of fatty acid and glycerin. When these glycerides of fatty acids are liquid at ordinary temperatures they are called oils, and those that are solids are called fats.

The grease content of wastewater can cause many problems in wastewater treatment plant processes. For example, high grease content can cause clogging of filters, nozzles, and sand beds (Gilcreas, Sanderson, & Elmer, 1953). Moreover, grease can coat the walls of sedimentation tanks and decompose and increase the amount of scum. Additionally, if grease is not removed before discharge of the effluent, it can interfere with the biological processes in the surface waters and create unsightly floating matter and films (Rowe & Abdel-Magid, 1995). In the treatment process, grease can coat trickling filters and interfere with the activated biosolids process, which, in turn, interferes with the transfer of oxygen from the liquid to the interior of living cells (Sawyer, McCarty, & Parking, 1994).

Carbohydrates, which are widely distributed in nature and found in wastewater, are organic substances that include starch, cellulose, sugars, and wood fibers; they contain carbon, hydrogen, and oxygen. Sugars are soluble while starches are insoluble in water. The primary function of carbohydrates in higher animals is to serve as a source of energy. In lower organisms, e.g., bacteria, carbohydrates are utilized to synthesize fats and proteins as well as energy. In the absence of oxygen the end products of decomposition of carbohydrates are organic acids, alcohols, as well as gases such as carbon dioxide and hydrogen sulfide. It should be pointed out that the formation of large quantities of organic acids can affect the treatment process by overtaxing the buffering capacity of the wastewater resulting in a drop in pH and a cessation of biological activity (Rowe & Abdel-Magid, 1995).

Detergents (surfactants) are large organic molecules that are slightly soluble in water and cause foaming in wastewater treatment plants and in the surface waters into which the effluent is discharged. Probably the most serious effect detergents can have on wastewater treatment processes is in their tendency to reduce the oxygen uptake in biological processes. In addition, according to Rowe and Abdel-Magid (1995), "detergents affect wastewater treatment processes by (1) lowering the surface, or interfacial, tension of water and increase its ability to wet surfaces with which they come in contact; (2) emulsify grease and oil, deflocculate colloids; (3) induce flotation of solids and give rise to foams; and (4) may kill useful bacteria and other living organisms" (p. 83). Since the development and increasing use of synthetic detergents, many of these problems have been reduced or eliminated.

8.3.3.2 Inorganic Substances

Several inorganic components are common to both wastewater and natural waters and are important in establishing and controlling water quality. Discharges of treated and untreated wastewater, various geologic formations, and inorganic substances left in the water after evaporation contribute to the inorganic load in water (Snoeyink & Jenkins, 1988). Natural waters dissolve rocks and minerals with which they come in contact. As pointed out earlier, many of the inorganic constituents found in natural waters are also found in wastewater. Moreover, many of these constituents are added via human use. These inorganic constituents include pH, chlorides, alkalinity, nitrogen, phosphorous, sulfur, toxic inorganic compounds, and heavy metals. Each of these is briefly discussed in the following.

When the *pH* of a water or wastewater system is considered, we are simply referring to the hydrogen ion concentration. This hydrogen ion concentration or pH is important because it affects chemical reactions. Many biological systems function in a narrow pH range of from about 6.5 to 8.5. Moreover, the equilibrium relationships in water/wastewater are strongly influenced by pH. In addition, many of the important properties of wastewater are due to the presence of weak acids and bases and their salts.

The wastewater treatment process is made up of several different unit processes (these will be discussed later). It can be safely stated that one of the most important unit processes in the overall wastewater treatment process is disinfection. The point is that pH has an effect on disinfection. This is particularly the case in regards to disinfection using chlorine. For example, with increases in pH, the amount of contact time needed for disinfection using chlorine increases.

In the form of the Cl⁻ ion, *chloride* is one of the major inorganic constituents in water and wastewater. Sources of chlorides in natural waters are (1) leaching of chloride from rocks and soils; (2) in coastal areas, salt water intrusion; (3) from agricultural, industrial, domestic and human wastewater; and (4) from infiltration of groundwater into sewers adjacent to salt water. The salty taste produced by chloride concentration in potable water is variable and depends on the chemical composition of the water. In wastewater, the chloride concentration is higher than in raw water because sodium chloride (salt) is a common part of the diet and passes unchanged through the digestive system. Metcalf & Eddy (1991) make the point that because conventional methods of waste treatment do not remove chloride to any significant extent, higher than usual chloride concentrations can be taken as an indication that the body of water is being used for waste disposal.

As was pointed out in an earlier discussion, *alkalinity* is a measure of the buffering capacity of water and in wastewater helps to resist changes in pH caused by the addition of acids. Alkalinity is caused primarily by chemical compounds dissolved from soil and geologic formations and is mainly due to the presence of hydroxyl and bicarbonate ions. These compounds are mostly the carbonates and bicarbonates of calcium, potassium, magnesium, and sodium. Wastewater is usually alkaline. Alkalinity is important in wastewater treatment because anaerobic digestion requires sufficient alkalinity to ensure that the pH will not drop below 6.2; if alkalinity does drop below this level, the methane bacteria cannot function. For the digestion process to operate successfully the alkalinity must range from about 1,000 to 5,000 mg/L as calcium carbonate (Rowe & Abdel-Magid, 1995). Alkalinity in wastewater is also important when chemical treatment is used, in biological nutrient removal, and whenever ammonia is removed by air stripping (Metcalf & Eddy, 1991).

In domestic wastewater, "nitrogen compounds result from the biological decomposition of proteins and from urea discharged in body waste" (Peavy et al., 1985, p. 195). In wastewater treatment, biological treatment cannot proceed unless *nitrogen*, in some form, is present. Nitrogen must be present in the form of either organic nitrogen (N), ammonia (NH_3), nitrite (NO_2), or nitrate (NO_3). Organic nitrogen includes such natural constituents as peptides, proteins, urea, nucleic acids, and numerous synthetic organic materials. Ammonia is present naturally in wastewaters. It is produced primarily by de-aeration of organic nitrogen-containing compounds and by hydrolysis of urea. Nitrite, an intermediate oxidation state of nitrogen, can enter a water system through use as a corrosion inhibitor in industrial applications. Nitrate is derived from the oxidation of ammonia.

Nitrogen data is essential in evaluating the treatability of wastewater by biological processes. If nitrogen is not present in sufficient amounts, it may be necessary to add it to the waste to make it treatable (Metcalf & Eddy, 1991). When the treatment process is complete, it is important to determine how much nitrogen is in the effluent. This is an important point because the discharge of nitrogen into receiving waters may stimulate algal and aquatic plant growth. These, of course, exert a high oxygen demand at nighttime, which adversely affects aquatic life and has a negative impact on the beneficial use of water resources.

Phosphorous (P) is a macronutrient that is necessary to all living cells and is a ubiquitous constituent of wastewater. It is primarily present in the form of phosphates—the salts of phosphoric acid. Municipal wastewaters may contain 10–20 mg/L phosphorus as P, much of which comes from phosphate builders in detergents. Because of noxious algal blooms that occur in surface waters, there is much interest in controlling the amount of phosphorous compounds that enter surface waters in domestic and industrial waste discharges and natural runoff (Metcalf & Eddy, 1991). This is particularly the case in the United States since approximately 15% of the population contributes wastewater effluents to lakes, resulting in *eutrophication* of these water bodies. Eutrophication leads to significant changes in water quality. This problem can be controlled by reducing phosphorous inputs to receiving waters (Hammer, 1986).

Sulfur (S) is required for the synthesis of proteins and is released in their degradation. The sulfate ion occurs naturally in most water supplies and is present in wastewater as well. Sulfate is reduced biologically to sulfide, which in turn can combine with hydrogen to form hydrogen sulfide (H_2S). H_2S is toxic to animals and plants. H_2S in interceptor systems can cause severe corrosion to pipes and appurtenances. Moreover, in certain concentrations, H_2S is a deadly toxin.

Toxic inorganic compounds such as copper, lead, silver, arsenic, boron, and chromium are classified as priority pollutants and are toxic to microorganisms. Thus, they must be taken into consideration in the design and operation of a biological treatment process. When introduced into a treatment process, these contaminants can kill off the microorganisms needed for treatment and thus stop the treatment process.

Heavy metals are major toxicants found in industrial wastewaters; they may adversely affect the biological treatment of wastewater. Mercury, lead, cadmium, zinc, chromium, and plutonium are among the so-called heavy metals—those with a high atomic mass. (It should be noted that the term, heavy metals, is rather loose and is taken by some to include arsenic, beryllium, and selenium, which are not really metals and are better termed toxic metals). The presence of any of these metals in excessive quantities will interfere with many beneficial uses of water because of their toxicity (Metcalf & Eddy, 1991). Urban runoff is a major source of lead and zinc in many water bodies. The lead comes from the exhaust of automobiles using leaded gasoline, while zinc comes from tire wear (Davis & Cornwell, 1991).

8.3.4 BIOLOGICAL CHARACTERISTICS OF WATER/WASTEWATER

Any specialist or practitioner who works in the water or wastewater treatment field must have some knowledge of the biological characteristics of water and wastewater. This knowledge begins with an understanding that water may serve as a medium in which thousands of biological species spend part, if not all, of their life cycles. Moreover, it is important to understand that, to some extent, all members of the biological community are water-quality parameters. This is the case because their presence or absence may indicate in general terms the characteristics of a given body of water.

The presence or absence of certain biological organisms is of primary importance to the water/wastewater specialist. These are, of course, the *pathogens*. Pathogens are the organisms that are capable of infecting or transmitting diseases to humans and animals. It should be pointed out that these organisms are not native to aquatic systems and usually require an animal host for growth and reproduction. They can, however, be transported by natural water systems. These waterborne pathogens include species of bacteria, viruses, protozoa, and parasitic worms (helminths). In the following sections a brief description of each of these species is provided.

8.3.4.1 Bacteria

The word *bacteria* (singular: bacterium) comes from the Greek word meaning "rod" or "staff," a shape characteristic of many bacteria. Bacteria are single-celled microscopic organisms that multiply by splitting in two (binary fission). In order to multiply they need carbon from carbon dioxide if they are autotrophs, from organic compounds (dead vegetation, meat, sewage) if they are heterotrophs. Their energy comes either from sunlight if they are photosynthetic or from chemical reaction if they are chemosynthetic. Bacteria are present in air, water, earth, rotting vegetation, and the intestines of animals. Gastrointestinal disorders are common symptoms of most diseases transmitted by waterborne pathogenic bacteria. In wastewater treatment processes, bacteria are fundamental, especially in the degradation of organic matter which takes place in trickling filters, activated biosolids processes, and biosolids digestion.

8.3.4.2 Viruses

A *virus* is an entity that carries the information needed for its replication, but does not possess the machinery for such replication (Sterritt & Lester, 1988). Thus, they are obligate parasites that require a host in which to live. They are the smallest biological structures known, so they can only be seen with the aid of an electron microscope. Waterborne viral infections are usually indicated by disorders with the nervous system rather than of the gastrointestinal tract. Viruses that are excreted by human beings may become a major health hazard to public health (Metcalf & Eddy, 1991). Waterborne viral pathogens are known to cause poliomyelitis and infectious hepatitis.

Testing for viruses in water is difficult because: (1) they are small in size; (2) they are of low concentrations in natural waters; (3) there are numerous varieties; (4) they are unstable; and (5) there are limited identification methods available (Rowe & Abdel-Magid, 1995). Because of these testing problems and the uncertainty of viral disinfection, direct recycling of wastewater and the practice of land application of wastewater is a cause of concern (Peavy et al., 1985).

8.3.4.3 Protozoa

Protozoa (singular: protozoan) are mobile, single-celled, complete, self-contained organisms that can be free-living or parasitic, pathogenic or nonpathogenic, microscopic or macroscopic. Protozoa range in size from two to several hundred microns in length. They are highly adaptable and widely distributed in natural waters, although only a few are parasitic. Most protozoa are harmless, only a few cause illness in humans—*Entamoeba histolytica* (amebiasis), and *Giardia lamblia* (giardiasis) being two of the exceptions. Because aquatic protozoa form cysts during adverse environmental conditions, they are difficult to deactivate by disinfection and must undergo filtration to be removed.

8.3.4.4 Worms (Helminths)

Worms are the normal inhabitants in organic mud and organic slime. They have aerobic requirements but can metabolize solid organic matter not readily degraded by other microorganisms. Water contamination may result from human and animal waste that contains worms. Worms pose hazards primarily to those persons who come into direct contact with untreated water. Thus swimmers in surface water polluted by sewage or stormwater runoff from cattle feedlots and sewage plant operators are at particular risk. The *Tubifix* worm is a common organism used as an indicator of pollution in streams.

CHAPTER 9

Water/Wastewater Treatment

Like many technical endeavors, the practice of water and wastewater treatment has benefitted from numerous developments in science and engineering. Water and wastewater treatment has been, traditionally, more of an art than a science. With a history of development founded in experience, treatment is rapidly incorporating fundamental scientific and engineering principles. Advances in design and operation have progressed with better understanding of these fundamental principles. (Jost, 1992, p. 94)

9.1 INTRODUCTION

WHEN the average person, whose house is connected to a public water supply system, goes to his or her kitchen sink to draw a glass of water from the tap, it is likely that little if any thought is given to the process involved with bringing this water to the house. However, in many locations, it is common knowledge that household potable water comes from some "managed" source; this knowledge is reinforced each time the monthly or bimonthly water bill arrives. The fact of the matter is that the average household occupant does not really give much thought to water drawn from the household tap; that is, not until something unusual occurs.

When we talk about "unusual occurrences," what are we really talking about? Well, for example, unusual occurrences include: those times when the tap is turned on and no water pours out; or when water does pour out of the tap but is colored by rust, dirt, or some other sources; and/or those times when the water drawn from the tap has a distinct, offensive taste and odor. These are unusual occurrences that tend to gain the water consumer's full attention; that is, either when there is a lack of water or when its quality is quite offensive.

The local public water utility is typically tasked with the responsibility of providing high quality potable water to each household within its jurisdiction. Moreover, this same utility is normally responsible for treating the water. Most groundwater and surface water sources used to supply a public water system contain some impurities. The typical water treatment process utilized in treating water is based on the types and concentrations of impurities present.

Obviously, water requires treatment to remove these impurities to protect public health. However, it should be pointed out that water treatment may also include changing the character of the water in order to improve the water's aesthetic qualities. Finally, water must also be treated in order to comply with drinking water regulations.

In water treatment conducted for public health-related reasons, the contaminants in drinking water that can cause sickness or death are removed or neutralized. These contaminants fall into the general categories of organic and inorganic chemicals and various types of microorganisms. It should be pointed out that these contaminants may cause sickness right away, as a result of only one drink of contaminated water, or the hazard may not be noticed immediately; that is, adverse health effects may occur only after years of exposure. Obviously, these contaminants must be removed from the water.

Chemicals may also be added to drinking water to benefit public health. The best example is fluoride. It is well documented that the main benefit of maintaining fluoride in drinking water is to reduce the incidence of cavities in children's teeth. For this reason, many states require fluoridation of municipal water supplies.

In water treatment conducted for aesthetic-related qualities, water is treated to make it more palatable to the consumer. When water has an unpleasant taste, odor, or color, it is unacceptable. There can be other problems with water for use in the home. For example, water with extreme hardness, with high levels of dissolved solids, and/or with the tendency to cause stains in laundry or on fixtures is also unacceptable.

Within economic constraints, water suppliers try to furnish the best possible water they can.

In water treatment for regulatory-related requirements, water is treated to comply with USEPA Primary Drinking Water Regulations. USEPA's drinking water regulations specify a *maximum contaminant level* (MCL) for most regulated contaminants. In the United States, if a water system exceeds any of these levels, the system must take steps to provide treatment to lower the level of contamination or change its source of water.

USEPA has also provided *secondary drinking water standards* that limit impurities that do not pose a known health threat. These secondary MCLs are not mandatory; instead, they are strongly recommended. One thing is certain, if these secondary limits are ignored, the public generally finds the water disagreeable and insists that corrections be effected ASAP.

9.2 WATER TREATMENT SYSTEMS

As was pointed out above, the primary purpose of a water treatment system is to bring raw water up to drinking water quality standards. The quality of the source water usually dictates the particular type of treatment process required to meet these standards. In some cases the water source may require only simple disinfection. Surface water will usually need to be filtered and disinfected. Groundwater, on the other hand, will often need to have hardness (calcium and magnesium) removed before disinfection.

As shown in Figure 9.1, a typical water treatment plant for surface water might include the following sequence of steps:

- *screening* to remove relatively large floating and suspended debris
- *mixing* the water with chemicals that encourage suspended solids to coagulate into larger particles of floc
- *flocculation*, which is the process of gently mixing the water and coagulant allowing the formation of large particles of floc
- *sedimentation* in which the flow is slowed enough so that gravity will cause the floc to settle
- *sludge processing* where the mixture of solids and liquids collected from the settling tank are dewatered and disposed of
- *disinfection* of the liquid effluent to ensure that the water is free of harmful pathogens
- *hardness removal* (This process is not usually part of the water treatment process where the source is surface waters. This is the case because surface waters seldom have hardness levels above 200 mg/L as $CaCO_3$. For groundwater, however, hardness levels are often as high as 1,000+ mg/L and must be treated to reduce hardness.)

9.2.1 SCREENING

Screening in water treatment is a preliminary pretreatment step. Screening is necessary to remove sticks, weeds, leaves, and other floating debris as well as rocks, gravel, sand, and other gritty substances. It is important to remove these materials to prevent downstream process equipment (pipes and pumps) from being damaged.

Manually and automatically cleaned *bar screens* and *wire-mesh screens* are the two basic types of screens used by water treatment plants. *Bar screens* are made of straight, vertical steel bars, welded at both ends to horizontal steel members. The screens are classified by the open distance between bars. A *fine* bar screen has spacing of approximately 1/16 to 1/2 in. A *medium* screen has spacing of 1/2 to 1 in. A *coarse* screen has spacing of

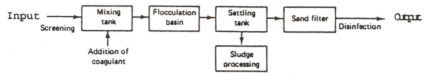

Figure 9.1 Typical water treatment process. Water softening may need to be added for groundwater. [Adapted from Masters (1991), p. 242.]

1/4 to 4 in. Bar screens are installed in the water influent stream at a slope ranging from 60 to 80° from the horizontal. This slope is important because it helps keep the screen from clogging between cleanings. In addition, the slope makes it convenient to rake debris up the screen face and onto the operating platform for drainage and disposal.

Automatically cleaned bar screens are available in a variety of models. Most automatic screens have some type of automatic rake assembly that works its way up the screen face, pushing debris to a drop-off point and onto a mechanical belt for deposit into a bin or hopper, which is then dumped for disposal.

Wire-mesh screens are commonly referred to as traveling water screens. These screens are constructed of fabric woven from corrosion-resistant wire-like materials such as stainless steel. The screen openings range from 1/60 to 3/8 in. in size. If the wire-mesh screen does not require frequent cleaning, it usually is of the manually cleaned variety. In larger operations, automatic, continuously cleaned wire-mesh screens are normally installed.

9.2.2 COAGULATION

After screening, raw water may still contain suspended particles of color, turbidity, and bacteria that are too small to settle in a reasonable time period, and cannot be removed by simple filtration. Since these fine particles and light solids often form stable suspensions, these suspensions must be destabilized to promote rapid settling under quiescent conditions using chemicals such as either mineral salts or synthetic organic polymers. Thus, the next step in the water treatment process is *coagulation*. Simply stated, coagulation is the process of adding and rapid mixing of chemical coagulants into the raw water. The object of coagulation is to alter these particles in such a way as to allow them to adhere to each other. Thus, they can grow to a size that will allow removal by sedimentation and filtration. Coagulation is a *chemical treatment process* that destabilizes particles in the size range of about $0.001-1$ μm (colloidal-sized particles) (Masters, 1991).

The exact way in which coagulants work is not clearly understood, but the intended purpose of the chemical coagulant is. The coagulant neutral-

izes the negative charge that causes the particles suspended in water to re-
pel each other. Thus, the coagulant allows particles to come together to
form larger particles that can be more easily removed from the raw water.

Table 9.1 provides a listing of commonly used coagulants. Topping this
list is aluminum sulfate (alum), which is the most commonly used chemical
coagulant. Ferric chloride, ferric sulfate, and polymers are also used.

Because alum is the most common coagulant used for water treatment, it
is important to understand the reactions involving alum. Thus, let's look at
these reactions.

(1) Alum added to raw water ionizes, producing Al^{3+} ions, some of which
    neutralize the negative charges on the colloids.
(2) Most of the aluminum ions, however, react with alkalinity in the water
    (bicarbonate) to form insoluble aluminum hydroxide $[Al(OH)_3]$.
(3) The aluminum hydroxide absorbs ions from solutions and forms larger
    particles (precipitate) of $Al(OH)_3$ and absorbed sulfate.
(4) Finally, the particles begin to collide and stick together (agglomerate)
    to form settleable floc particles (Masters, 1991; AWWA, 1995b).

Once coagulants have been added to raw water, it is necessary to provide
rapid agitation to distribute the coagulant evenly throughout the water. This
rapid mixing is accomplished in a chamber that has rapidly rotating mixing
paddles. This rapid mixing process typically takes about one minute.

9.2.3 FLOCCULATION

After coagulation, *flocculation* follows in a tank where slow mixing of
the chemicals with water assists in building up particles of *floc*. It should be
pointed out that this "tank" does not necessarily have to be a separate, independent vessel as shown in Figure 9.1. In some systems, coagulation, flocculation, and sedimentation can be combined into a single unit.

TABLE 9.1. Common Coagulation Chemicals.

Common Name	Comments
Aluminum sulfate	Most common coagulant in the U.S.
Ferric chloride	May be more effective than alum in some applications
Ferric sulfate	Often used with lime softening
Aluminum polymers	Includes polyaluminum chloride and polyaluminum sulfates
Sodium aluminate	Used with alum to improve coagulation
Sodium silicate	Ingredient of activated silica coagulant aids

Source: Adapted from AWWA, 1995b, p. 57.

In the flocculation stage, slow, gentle agitation takes place in a chamber for approximately one-half hour. During agitation, smaller particles of floc are transformed into larger aggregates of visible floc.

9.2.4 SEDIMENTATION

After flocculation, the water flows through a sedimentation basin, or clarifier, where the primary function of the sedimentation process is accomplished. That is, the water is prepared for effective and efficient filtration. This preparation is accomplished in a large circular or rectangular concrete tank designed to hold the water for a long enough time (usually 1 to 10 hours) to allow most of the suspended solids to settle out. Solids that collect on the bottom of the tank are removed manually by periodically shutting down the tank and working out the collected sludge, or the tank may be continuously and mechanically cleaned using a bottom scrape. The effluent from the tank then moves on to the next step in the process: *filtration.*

9.2.5 FILTRATION

The purpose of *filtration* is to remove almost all of the suspended matter that remains. This suspended material that must be removed from surface water may include floc from the coagulation, flocculation, and sedimentation processes. In addition, microorganisms and precipitate are also filtered out.

Probably the most widely used filtration unit process is called a *rapid-sand filter.* This filter system consists of a layer of carefully sieved sand on top of a bed of graded gravel. It is interesting to note that the pore openings between grains of sand are often greater than the size of the floc particles that are to be removed. The point is that much of the actual filtration is accomplished by adsorption, continued flocculation, sedimentation in the pore spaces, as well as by simple straining. When the filter becomes clogged with particles, the filter is shut down and cleaned by forcing water backwards (backwashing) through the sand for about 15 minutes. After the cleaning, which occurs roughly once a day, the sand settles back in place and operation resumes.

9.2.6 DISINFECTION

Disinfection is the final water treatment process; it is used to destroy or inactivate pathogenic (disease-causing) organisms. Diseases caused by pathogenic organisms are called waterborne diseases; the more common ones are summarized in Table 9.2.

TABLE 9.3. Common Waterborne Diseases.

Waterborne Disease	Causative Organism	Source of Organisms in Water
Gastroenteritis	*Salmonella*	Animal or human feces
Typhoid	*Salmonella typhosa*	Human feces
Dysentery	*Shigella*	Human feces
Cholera	*Vibrio comma*	Human feces
Infectious hepatitis	Virus	Human feces (shellfish grown, polluted waters)
Amoebic dysentery	*Entamoeba histolytica*	Human feces
Giardiasis	*Giardia lamblia*	Animal or human feces
Cryptosporidiosis	*Cryptosporidium*	Human and animal feces
Legionellosis	*Legionella pneomophilia*	—

Chlorination is the most common form of disinfection employed in the United States. When properly operated, the chlorination process is a safe, practical, and effective way to destroy disease-causing organisms. Chlorination chemicals such as chlorine gas, sodium hypochlorite, or calcium hypochlorite are the most common agents used in disinfection. Chlorine is a powerful oxidizing agent that is easy to use, inexpensive, and reliable. It should be pointed out, however, that although chlorination is completely effective against bacteria, its effectiveness is less certain with protozoal cysts and viruses.

There are a few disadvantages in using chlorine for disinfection. The first disadvantage is the potential formation of trihalomethanes (THMs), some of which are carcinogens. THMs are created when chlorine combines with natural organic substances, such as decaying leaves, which may be present in the water itself. Another disadvantage in using chlorine is the impact of OSHA's Process Safety Management Standard (PSM), 29 CFR 1910.199. Under PSM, facilities using more than 1,500 pounds of chlorine must comply with the 16 elements of the standard. The PSM Standard is designed to lessen the risk of a catastrophic release of deadly chlorine gas into the environment. The intent of PSM is difficult to argue against. However, facilities who are required to implement PSM are required to expend a substantial outlay of capital to ensure full compliance. Complicating the situation is USEPA's new Risk Management Program (RMP). Facilities who use more than 2,500 pounds of chlorine in their process may not only be required to implement all or part of PSM but will also be required to conduct an offsite consequence analysis of potential chemical spills, and then make this information available to the public. Obviously, plant managers who must comply with RMP (effective date: June 1999) are somewhat "antsy" about having to inform their neighbors that the plant contains enough deadly chlorine gas that could, in a worst-case scenario, affect the neighborhood—or even worse. Thus, PSM and RMP require decisions to be made. Specifically, will the plant continue to use chlorine in its process or substitute with a process that is not regulated by PSM and/or RMP? At this time it is too early to predict which direction chlorine users are going to take (Spellman, 1997c).

9.2.7 SOFTENING

The reduction of hardness, or *softening,* is an additional process commonly practiced in hard-water areas. Softening can be accomplished by the consumer at the point of use or by the water utility. As a rule of thumb, when water contains more than 150 mg/L hardness, it should be softened at the water-treatment plant. At the water-treatment plant, chemical precipitation and ion exchange are the softening processes most commonly used.

In the *chemical precipitation process,* softening is accomplished by converting calcium hardness to calcium carbonate and magnesium hardness to magnesium hydroxide. This can be accomplished by the lime-soda ash process or by the ion exchange process.

The *lime-soda ash process* has the advantages of reducing the total mineral content of the water, removing suspended solids and precursors of THMs, removing iron and manganese, and reducing color and bacterial numbers. The process, however, produces large quantities of sludge, requires careful operation, and, if pH is not properly adjusted, may create problems with filter media and within the distribution system (McGhee, 1991).

The *ion exchange process* functions by charging a resin with sodium ions and allowing the resin to exchange the sodium ions for calcium and/or magnesium ions. Common resins include natural or man-made minerals that will collect from a solution certain ions and exchange these ions. These resins are placed in a pressure vessel. A salt brine is flushed through the resins. The sodium ions in the salt brine attach to the resin; the resin is now "charged." Once charged, water is passed through the resin, the resin exchanges the sodium ions attached to the resin for calcium and magnesium ions, thus removing them from the water.

The advantage of the ion exchange process, used in water softening, is that it removes all or nearly all of the hardness. The disadvantage is that sodium is added to the water. This additional sodium results in the water being more corrosive. Moreover, the additional sodium may increase the health risk to those consumers with high blood pressure.

9.3 WASTEWATER TREATMENT

Simply stated, at the present time, proper management of wastewater is a necessity, not an option. However, this was not always the case. For example, in the U.S., the treatment of wastewater did not receive much attention a century ago because the extent of the nuisance caused by the discharge of untreated wastewater into the relatively large bodies of water was not severe. However, during the early 1900s, nuisance and health conditions reached such deplorable levels that the need for more effective means of wastewater management was recognized. After finally coming to the realization that something had to be done to clean up the waste stream, another problem became apparent. That is, land space was no longer readily available for disposal of untreated wastewater, especially in larger settlements (cities). Because of this problem the planning, design, construction, and operation of a higher level (more intensive methods) of wastewater treatment evolved (Metcalf & Eddy, 1991).

There are several problems related to untreated wastewater: (1) if untreated wastewater is allowed to accumulate, the decomposition of the organic materials it contains can lead to the production of offensive odors and gases; (2) untreated wastewater contains numerous pathogenic microorganisms; (3) wastewater contains nutrients that can stimulate growth of aquatic life; and (4) wastewater may also contain toxic compounds. For these reasons, wastewater treatment is imperative.

The purpose of wastewater treatment is to convert the components in raw wastewater into a relatively harmless final effluent for discharge to a receiving body of water and to safely dispose of the biosolids produced in the process. The planning, design, construction, and operation of wastewater treatment facilities is a complex problem. Today it can be safely stated that wastewater treatment is no longer an "art"; instead, it has become a "science."

The primary purpose of wastewater treatment is to provide the principal means of improving water quality in point-source discharges. Thus, the typical wastewater treatment plant is designed (1) to prevent pollution of water supplies and contamination of aquatic organisms intended for human consumption (e.g., shellfish); (2) to prevent the pollution of recreational areas; (3) to prevent nuisance, unsightliness, and unpleasant odors; (4) to prevent human wastes coming into contact with humans, animals, and foods or being exposed on the ground surface accessible to children and pets; and (5) to comply with standards for groundwater and surface waters related to wastewater disposal and water pollution control. Protection of land and water resources is a national policy (in many countries) and every effort must be made to prevent their pollution by improper treatment and disposal of sewage and industrial wastewater.

Before beginning a brief description of the wastewater treatment process it is desirable to define some commonly used terms.

TABLE 9.3. Typical Range of Composition of Untreated Domestic Wastewater.

Constituent	Concentration (mg/L)
5-day biochemical oxygen demand (BOD$_5$)	100–300
Chemical oxygen demand (COD)	250–1000
Total dissolved solids (TDS)	200–1000
Suspended solids (SS)	100–350
Total Kjeldahl nitrogen (TKN)	20–80
Total phosphorous (as P) (TP)	5–20

Source: Adapted from Davis and Cornwell, (1985). p. 523.

Domestic sewage is the used water from a home or community; it includes toilet, bath, and kitchen-sink wastes. Sewage from a community may include industrial and commercial wastes, groundwater, and surface water. Hence the more inclusive term *wastewater* is also in general usage. The terms are used interchangeably. Normal domestic sewage will average about 99.9% water. The characteristics of the remaining portion (solids) vary somewhat from location to location, with the variation depending primarily on inputs from industrial facilities that mix with the somewhat predictable residential flows. Table 9.3 shows the typical range of composition (generalized) of untreated domestic wastewater. The strength of wastewater is commonly expressed in terms of 5-day biochemical oxygen demand (BOD), suspended solids, and chemical oxygen demand (COD), which are listed in Table 9.3.

Privy or one of its modifications is the common device used when human waste (excreta) is disposed without the aid of water. When excreta is disposed of with water, a *water-carriage* sewage-disposal system is used; generally all other domestic wastes are included. When storm water and domestic sewage enter a sewer it is called a *combined sewer.* If domestic sewage and storm water are collected separately, in a *sanitary sewer* and in a *storm sewer,* the result is a *separate sewer system.* A *sewer system* is a combination of sewers and appurtenances for the collection, pumping, and transportation of sewage, sometimes called *sewerage;* when facilities for treatment and disposal of sewage, known as the *sewage* or *wastewater treatment plant,* are included the reference would be to a *sewage works* (Salvato, 1982, pp. 380, 382).

9.3.1 HOW DOES WASTEWATER TREATMENT WORK?

A wastewater treatment plant cleans used water through a series of physical, mechanical, biological, and chemical processes. Treatment methods that rely on biological and chemical processes are called *unit processes*. Biological unit processes involve microbial activity that is responsible for degradation of organic matter and removal of nutrients (Metcalf & Eddy, 1991). Chemical unit processes include disinfection, adsorption, and precipitation. Physical unit processes include screening, sedimentation, filtration, and flotation.

Wastewater treatment begins with physical processes (unit operations) and basically follows four steps:

(1) *Preliminary treatment:* The objective of this unit operation is to remove debris and coarse materials that may clog piping and equipment in the plant.

(2) *Primary treatment:* Treatment is brought about by unit operation such as screening and sedimentation.

(3) *Secondary treatment:* Biological unit processes such as activated biosolids, trickling filter, oxidation ponds, etc. and chemical unit processes such as disinfection are used to treat wastewater. Removal of nutrients also generally occurs during secondary treatment.

(4) *Advanced or tertiary treatment:* Unit operations and chemical unit processes are used to further remove BOD, pathogens and parasites, nutrients, and toxic substances (Bitton, 1994).

The following sections present a brief description of the three types of wastewater treatment: primary, secondary, and tertiary.

9.3.2 PRIMARY TREATMENT

Primary treatment plants use unit operations or physical processes, such as screening and sedimentation, to remove a portion of the pollutants that will settle, float, or that are too large to pass through simple screening devices. This is followed by disinfection. Primary treatment typically removes about 35% of the BOD and 60% of the suspended solids. Primary treatment removes the most visibly objectionable substances and provides some degree of safety by disinfection. The problem is that the effluent still contains enough BOD to cause oxygen depletion problems and enough nutrients, such as nitrogen and phosphorous, to accelerate eutrophication.

In regards to the removal of pathogens, primary treatment removes the heavier and larger organisms, such as cysts, protozoa, and the eggs of helminths. In addition, primary treatment may remove organic matter associated with the microorganisms. Primary treatment is not efficient in the reduction of bacteria and viruses in wastewater (Clark et al., 1961).

Some chemical constituents, such as metals, some organic nitrogen, organic phosphorus, and heavy metals are removed. Heavy metals (e.g., arse-

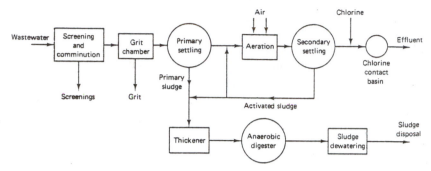

Figure 9.2 Typical wastewater treatment process. [Adapted from Masters (1991), p. 242.]

nic, barium, copper, and cadmium) are removed by sedimentation (precipitation), that are precipitated as hydroxides by the addition of lime or caustic soda to a pH of minimum solubility (Asano et al., 1985; Eckenfelder, 1989).

9.3.3 SECONDARY TREATMENT

In secondary treatment, treatment shifts from unit operations to unit processes. After primary treatment, secondary treatment submits wastewater to the additional step of biological (bacterial) degradation that takes place in the aeration tank (see Figure 9.2). This step brings the total BOD and suspended solids removed to approximately 90%.

Secondary treatment of wastewater took a giant leap forward because of the Clean Water Act (CWA) of 1977. This act required at least secondary treatment for all publicly-owned treatment works (POTWs) by stipulating that such facilities provide at least 85% BOD removal for those discharging into freshwater bodies. What the CWA stipulates is production of an effluent requirement of 30 mg/L for both 5-day BOD and suspended solids (monthly average).

The physical processes that make up primary treatment are augmented in secondary treatment with processes that involve the microbial oxidation of wastes. This biological treatment process is analogous to putting a stream into a "box" and letting the self-purification process that occurs naturally in streams take place within the aeration basins; thus, mimicking nature's way of using microorganisms to oxidize the organics (Spellman, 1996b). This man-made treatment process does have a considerable advantage over nature, however. The advantage gained is this system allows man to *control conditions* in the treatment plant itself—leading to optimization in treatment operations.

The basic ingredients required for aerobic secondary biologic treatment are microorganisms, good contact between these organisms and the organic material, the availability of oxygen, and the maintenance of other environmental conditions such as temperature and sufficient time for the organisms to work (Davis & Cornwell, 1991). There are three commonly used approaches that have been used in the past to meet these basic needs: (1) *trickling filters,* (2) *activated biosolids,* and (3) *oxidation ponds or lagoons.*

In *trickling filters,* contact between organics and microorganisms is optimized by passing the wastewater over a film of biomass attached to solid surfaces (rocks or synthetic media) (Peavy et al., 1985). A variation of the trickling filter attached growth concept is provided by the *rotating biological contactor* (RBC). An RBC consists of a series of closely spaced, circular plastic disks that are attached to a rotating horizontal shaft. Almost one-half of each disk is submersed in a tank containing wastewater. The biomass film that grows on the surface of the disks moves into and out of the wastewater as the RBC rotates. While the organisms are submerged, they absorb organics; while they are out of the wastewater, they are supplied with needed oxygen (Masters, 1991).

In the *activated biosolids process,* a mixture of wastewater and biosolids (microorganisms) is agitated and aerated. The biosolids are subsequently separated from the treated wastewater and returned to the aeration process as needed (Davis & Cornwell, 1991).

Oxidation ponds or *lagoons* have been used to treat wastewater for many years, particularly as wastewater treatment systems for small communities (Benefield & Randall, 1980). An oxidation pond consists of a large, shallow earthen basin in which wastewater is retained long enough for natural purification processes (decomposition by microorganisms) to provide the necessary degree of treatment. The conditions are similar to those in an eutrophic lake (Masters, 1991).

Bacterial concentrations are greatly reduced by secondary treatment processes. Some secondary treatment processes claim a removal efficiency of more than 90% for the coliform indicator organisms, which, obviously, should relate to the removal of pathogens (USEPA, 1992). In regards to removal of chemical constituents, secondary treatment removes metals through the absorption of dissolved metals or fine particulate metals onto the biosolids flocs. Moreover, under normal operating conditions, the trace elements concentrations are reduced by about 85% (Chen et al., 1974).

9.3.4 TERTIARY TREATMENT

Tertiary treatment (advanced treatment) involves a variety of different techniques to remove dissolved pollutants left after primary and secondary treatments. The tertiary treatment of wastewater removes phosphorous and nitrogen that could increase aquatic plant growth, as well as to more completely reduce BOD and trace metals.

CHAPTER 10

Water Use

All rivers run into the sea, yet the sea is not full; into the place whence the rivers come, thither they return again. (Ecclesiastes 1:7)

10.1 INTRODUCTION

WE cannot live without water. This is the case even though water does not form part of the organic structure of any organism. Moreover, water does not provide energy for any organism and has no nutritional value in the accepted sense. In regards to water content, "it varies very considerably between organisms, between different parts of the same organism and in the same organism at different times" (Bradbury, 1991, p. 16).

If the above statements about water are true (and they are), then what is the big deal about water? Why is it that we cannot live without it?

Water is critical to life because all biochemical processes (processes essential to life) occur in water. Water is also critical to life because it is the material (medium) that transports substances around the body and removes unwanted substances (wastes). Water also provides surface cooling through dissipation of heat that occurs when water evaporates from the organism.

The bottom line: Like food, water is important (absolutely essential) for human life.

The average person needs about 1.5 liters of water daily to survive.

After having read this statement you might think that the basic problem related to water is a lack of drinking water; however, this is not the case. The fact is few people die of thirst. Instead, the problem is to obtain a sufficient supply of safe drinking water and adequate sanitation services. This lack of water supply and sanitation facilities is very great, particularly in lesser-developed countries. This problem is magnified when you consider that only 51% of the people in developing countries have access to safe water.

Modern civilization, more than any preceding it, places heavy demands on water.

197

TABLE 10.1. Freshwater Usage in the U.S. (10^9 m³/year).

Usage	Withdrawals	Consumption	Returns
Irrigation	233	122	101
Municipal	40	10	30
Manufacturing/minerals	80	11	69
Thermal electric power	123	2	121
Totals	467	145	322
Groundwater withdrawals	113		
Surface water withdrawals	354		

Source: Adapted from U.S. Water Resources Council (1978), p. 17. Totals may not add due to rounding.

In regards to water use, there are two common measures of water use: *withdrawal water use* and *consumption water use.* Water is withdrawn when it is taken from a surface or ground source and conveyed to the place to use. Water is consumed when, after withdrawal, it is no longer available for reuse in the local area because of evaporation, contamination, storage in living plant and animals, and seepage into the ground. About 25% of our water withdrawals are consumed, with most of that consumption resulting from evapotranspiration of irrigation water.

It is important to note that water that is withdrawn and used for some purpose can usually be returned and reused again, provided it is not too polluted. Thus, it is important to distinguish between water *withdrawals,* water *returns,* and *consumptive uses* (Masters, 1991). The point is water that is withdrawn from a source is either consumed or returned:

$$\text{Withdrawals} = \text{Consumption} + \text{Returns}$$

Table 10.1 shows freshwater use in the United States and includes amounts withdrawn, consumed, and returned. About 75% of U.S. freshwater needs are supplied from surface water and the remaining 25% is derived from groundwater. It is important to note that water distribution is very unevenly divided across the United States; in some locations there is plenty of water and in others not enough (this problem will be discussed in more detail later).

Before beginning a detailed discussion on how and where water is used and various water supply problems, a few general points are made about the "state-of-water" today.

- If water is used wisely and efficiently, adequate quantities of water are available to satisfy worldwide demands through the year 2000. On the other hand, shortages of water in some areas will occur if poor management, lack of adequate conservation, pollution, and rapid local increases continue.

- Only about half the people have access to safe drinking water in developing nations. Moreover, some 10 million deaths each year, worldwide, result from waterborne intestinal diseases.
- Surface and underground water supplies, in industrial nations, are being polluted by industrial and municipal wastes, and by surface runoff from urban and agricultural areas.
- In China, India, and the United States, heavy demands for water by agriculture, industry, and municipalities are rapidly depleting groundwater supplies.
- Proper protection and management of watershed areas, creation of incentives for conservation, and legislation that encourages water recycling are required to ensure adequate water supplies.
- Ensuring adequate water supplies is important; however, it is even more important to ensure acceptable water quality. This can only be accomplished if programs are implemented to reduce the generation of solid, liquid, and airborne wastes from industrial plants, mining and smelting operations, electric power production, cities and towns, and agriculture.

Unless we are thirsty, the natural tendency is to take water for granted. This is not that hard to understand when the prevailing mind-set thinks of water as nothing more than "just water." The point is that since water is renewable, unlike fossil fuels and soil, we just don't worry about it.

Why should we worry about water? Isn't water just about everywhere?

Well, it is true that scientists know that if water is properly used and carefully conserved, the global hydrological cycle can meet current and anticipated freshwater needs on a sustainable basis. It is also true, however, that problems of freshwater supply and water quality are of immediate, fundamental, and profound importance to all people. Again, the problem is that unless you are thirsty, water is not something to worry about, right?

When you get right down it, that is, when you get right down to whether we should worry about water, we have to consider the situation, the trend, the future. For example, when you consider that population growth and rising requirements for energy and food are placing greater demands on both the quantity and quality of freshwater supplies, then water is something to worry about. When you consider that the total worldwide water use more than tripled between 1950 and 1980, then water is something to worry about (Postel, 1985). Further, when you consider that in developing countries, problems of water pollution and depletion are dominant, then water is something to worry about.

In order to gain an appreciation for and an understanding of the slowly intensifying wind at our backs that is insidiously propelling the realization about the growing problem concerning the "water issue," the following three sections present major aspects of the water issue: (1) water supply and demand (global), (2) water supply problems, and (3) water resources management. Chapter 11 presents a discussion of another important (probably the most important) aspect of the water issue: water pollution. Then, appropriately and fittingly, Chapter 12 discusses water reuse.

10.2 WATER SUPPLY AND DEMAND (GLOBAL)

As previously mentioned, water covers three quarters of the earth's surface, but more than 97% of the earth's water is salt water in the oceans and seas, and less than 3% is fresh water. Of the latter, approximately 78% is frozen in polar ice caps and glaciers, 21% is groundwater, and the remaining 1% is in lakes, streams, rivers, plants, and animals (WRI & IIED, 1988).

History shows that the world's streams, rivers, and lakes have provided important resources and services, including drinking water, water for washing, agriculture, transportation, energy production, waste disposal, and recreation (WRI & IIED, 1988).

In regards to water demand, earlier it was pointed out that demand is on the rise. This rise in demand is not gradual, however. In the last 20 to 30 years worldwide population has increased exponentially in comparison to periods prior to this period. The point is that when population increases, so does the demand for water.

10.3 WATER SUPPLY PROBLEMS

The first water supply problem that comes to mind is the one that is most obvious and has the most impact on plants, animals, and man. The problem we are talking about here, of course, is *too little water*. Few experts in the field would argue against the fact that the availability of adequate supplies of freshwater is the most serious long-range problem confronting the United States and the rest of the World. However, the nonexpert who lives in the Pacific Northwestern part of the United States or in Norfolk, Virginia, this may seem an implausible and insignificant statement (scare tactics). To these folks this is the case, of course, because these two geographic areas have plenty of potable water and nearly the same amount of annual rainfall per year. (Did you know that Norfolk actually has a higher annual rate of precipitation than Seattle?) However, others, especially those who reside in the 80+

arid and semiarid countries (accounting for 40+% of the world's population) have no problem understanding that too little water is a very real problem. With a history of serious periodic droughts and considerable difficulty with growing enough food to support their populations, residents of arid and semiarid regions understand what the implications (life-threatening in some instances) of having too little water are.

Water shortages can be the result of several factors, including limited supplies, heavy demands, and inefficient use.

It is likely that future water shortages will limit growth in agriculture and industry, and could jeopardize health, nutrition, and economic development.

It should be pointed out that the "too little water" problem extends beyond the arid and semiarid regions. For example, in many lesser-developed countries, poor people must spend a considerable part of their waking hours fetching water, often from polluted sources. To obtain water, many residents of lesser-developed countries walk 10 to 15 miles a day, carrying heavy water-filled containers on their return trip.

There is little doubt that droughts can cause all sorts of problems. It should be noted, however, that drought is one thing, but rapid population growth coupled with poor land use is another. When you combine drought with increased numbers of thirsty people who have poorly managed their land, the adverse effects are exponential. In order to survive, many residents of lesser-developed countries must resort to cutting more trees, growing more crops at more erosion-prone elevations, cultivating poor soils, and allowing their livestock to roam free and overgraze whatever grassland area there might be. All of these actions work to further degrade the land, which in turn reduces the amount of rainfall absorbed and slowly released by soils and vegetation. The end result, in many cases, is decertification.

Along with natural and man-made causes of drought (environmental degradation), the question becomes: What are the other causes of water shortages? Answer: growing populations and increasing use of water in agriculture (more people require more food).

World Resources Institute (WRI) and International Institute for Environment and Development (IIED) (1987) point out that worldwide irrigation for agriculture accounts for about 73% of water use; 21% goes to industry; and the other 6% to domestic use. According to Masters (1991), in the U.S., "we pipe, pump, or divert an amount of fresh water and saline water equivalent to about 40% of our runoff to meet our domestic, industrial, cooling water, rural, and irrigation needs" (p. 104). It is interesting to note that agriculture accounts for more than 80% of total consumptive water use in the United States and that this use is very inefficient; most of the water is lost by evaporation and/or seepage into the ground (Pimental, 1989). Irrigation, worldwide, is only about 37% efficient (Postel, 1984).

In 1980, a survey conducted by Worldwatch Institute pointed out that to-

TABLE 10.2. Estimated Water Withdrawals in Selected Countries, Total, 1980.

Country	Total (billion liters/day)
United States	1,683
Canada	120
Soviet Union	967
Japan	306
Mexico	149
India	1,058
United Kingdom	78
Poland	46
China	1,260

Source: Adapted from Postel, S. (1984), p. 16.

tal and per capita U.S. water withdrawals from surface and groundwater supplies were substantially greater than withdrawals for nine other nations. Daily U.S. withdrawals were 7,200 liters per person, compared with 4,800 liters for Canada and only 1,400 for the United Kingdom (see Table 10.2) (Postel, 1985). In the Southwestern Soviet Union, water withdrawals for irrigation are primarily responsible for a 40% decrease in area of the Aral Sea, formerly, in area, the world's fourth largest lake (Micklin, 1988).

Postel (1985) points out that one fifth of the water pumped out of the ground in the United States each year is nonrenewable. Rapid development and increasing use of irrigation, especially in the West, have caused extensive water withdrawals from streams and rivers in the Rio Grande Basin and Lower Colorado. Demand is so intense that some states must bring in water from other areas or pump more from groundwater reservoirs. The problem is that water is being withdrawn more rapidly than it can be replenished. Consider, for example, the Ogallala aquifer; it is already 50% depleted while being required to supply one fifth of all U.S. irrigated cropland (WRI & IIED, 1986).

Damage to a drawn-down aquifer is serious; it can be irreversible. The damage is irreversible in the sense that groundwater overdraft can cause settling of the land (subsidence) as water is removed. Weber (1988) points out that a collapsed aquifer can never be replenished.

In 1940, U.S. water withdrawals per day were 140 billion gallons as compared to 450 billion gallons withdrawn in 1985 (U.S. Bureau of Census, 1987).

In many lesser-developed countries, grazing and cultivation practices in steep, high-rainfall zones and deforestation of watersheds for commercial timber or fuelwood use reduce the soil's capacity to absorb rainfall, which increases flooding downstream and reduces the amount of water available during dry seasons (Quigg, 1976). This is only part of the problem, however. Soil eroded from uplands causes siltation of reservoirs used for water storage.

Water shortages lead to other problems. For example, when water shortages occur, international rivers and lakes become the focus of growing tension. Over 200 of the largest river systems are shared by 2 to 10 nations. Obviously, in order to prevent conflict, a coordinated water resource policy is needed for each of the affected nations.

After reading the above information, the reader might be confused about the other problem with the world's water supply; that is, *too much water.* Too much water really is a problem. Consider, for example, the problem faced by India each year. India receives enough precipitation on an annual basis but most of it is delivered at one time of year. 90% of India's annual precipitation falls between June and September—the monsoon season. The problem is not only with periodic flooding; the downpour runs off the landscape so rapidly that most of it cannot be captured and used.

During any given year, major flood disasters affect millions of people, kill thousands of them each year, and cause tens of billions of dollars in property damage. The irony is that floods are normally classified as natural disasters, but the fact is that human beings have contributed to the sharp rise in flood deaths and damages by removing water-absorbing soil and vegetation through deforestation, cultivation of marginal lands, overgrazing, and mining. Adding to the flooding problem is the massive movement toward urbanization that is now taking place. Urbanization increases flooding by replacing vegetation with parking lots, airports, highways, shopping centers, homes, office buildings, and numerous other structures.

Catastrophic flooding events, including death and extensive damage, have also increased because many poor people in lesser-developed countries have little choice but to live on land subject to severe flooding; many of these people insist that the benefits of living in flood-prone areas outweigh the risks. Many urban areas and croplands throughout the world are located on flat areas (floodplains) along rivers subject to periodic flooding. An example of this trend and its tragic consequences was clearly demonstrated by the 500-year flooding event that occurred in the Red River area of the upper Midwestern U.S. in early spring of 1997. The sad part of this devastating event goes far beyond the damage to property and human suffering. Indeed, the sad reality is that after the flood waters receded, many of the affected residents rebuilt their homes and their towns, several at significant taxpayer expense—on that same floodplain.

In the United States, several billion dollars have been spent in the last 70 years by the U.S. Army Corps of Engineers, the Bureau of Reclamation, and the Soil Conservation Service on flood-control projects such as channelization (straightening stream channels), dredging streams, and building dams, levees, reservoirs, and seawalls. Despite these efforts property damage (measured in billions of dollars) from floods in the United States has increased significantly in the last 25 years.

There are two significant problems with man trying to straighten stream channels, dredging streams, building dams, levees, and reservoirs: (1) Mother Nature and (2) these projects tend to stimulate development on flood-prone land. There is no question that man has made great strides in both science and technology. For example, technology has allowed us to tap vast underground water reservoirs and drain too many of them at rates that far exceed any hope for natural replenishment. In other cases, we drain and divert rivers for local concerns, oblivious to long-term ecological needs. The point is that there are some things that man cannot do or should not attempt to do. Specifically, we cannot prevent Mother Nature from having her way with us—and with everything else. When we attempt to gain the upper hand on Mother Nature, we generally lose. And even when we seemingly win one on Mother Nature, have we really won? What about the long-term consequences? Can any of us predict with perfect accuracy what our actions today will reap tomorrow—or a hundred years from now—a thousand? Unfortunately, many of man's well-intentioned activities cause people to be injured and cost billions of dollars in property damage.

There are other problems or adverse effects generated from water development projects, such as dams, canals, and stream channelization. These activities can seriously degrade water quality, spread waterborne diseases, destroy farmland, ruin wetlands, and contribute to species extinction through habitat destruction.

Instead of fighting Mother Nature, we should assist her. In this regard, there are several steps we can take in attempting to establish effective flood prevention/reduction methods: (1) we should replant vegetation in disturbed areas to reduce runoff; (2) we should build ponds in urban areas to retain rainwater and release it slowly to streams and rivers; and (3) we should install storm water systems to divert rainwater. The ultimate step, of course, is to clearly identify floodplains and prevent, through laws and zoning regulations, their use for urban development. This is an important point, especially when we consider that much of the growing population in developing countries is moving out of rural areas. Moreover, it is really not an exercise in over reaching to estimate that by the century's end, almost half of humanity will live in cities. Obviously, providing water for such enormous population centers will require great improvements in the future management of water resources.

You are correct. The above recommendations are easier said than done. No doubt about it.

How does that old saying go? We need to study history so that we will not repeat our past mistakes. Isn't this what we have been talking about here?

Another water supply problem has to do with *water that is in the wrong place*. Some countries have sufficient annual precipitation and very large rivers, which carry much of the runoff. The problem is that these rivers are located far from agricultural and population centers where the water is needed. For example, consider South America where average annual runoff is the largest in the world, but 60% of it flows through the Amazon River where few people live.

So, what does all this mean? What it means is that we can no longer treat water as an unlimited resource that is provided as cheaply as possible and in the quantity we desire. However, if we continue to view water in this way, critical deficiencies in quantity and quality of available water will result. Increased demands on water are being driven by population growth and rising requirements for energy and food.

The bottom line: To meet the demand, for the present and the future, nations must practice more efficient water management, introduce reuse, prevent pollution, and promote water conservation (Myers, 1984).

10.4 WATER RESOURCE MANAGEMENT

Statement: We cannot increase the earth's water; what we have now is what we are going to have—forever.

If we accept the above statement as fact, then it is apparent that we need to better manage the water that we have to reduce the impact and spread of water resource problems. In this effort, we have but two choices available to us: (1) increase the usable supply; and (2) decrease unnecessary loss and waste. Obviously, for any water resource management plan to be effective, a combination of these two approaches should be effected.

However, it should be pointed out that combining these two approaches into one concerted effort toward effective water management is no easy task. There are problems that must be overcome. For example, in the United States water has been managed by three different, separate interests: (1) water supply, (2) wastewater treatment and collection, and (3) storm and flood water management. Each of these separate entities has its own goals, procedures, and constituencies. The point is, at the present time, there is no single, overall strategy for managing water resources. When you think about this situation, it seems strange, it doesn't fit our perception or expectations. We have to ask ourselves a question: How many Fortune 500 companies operate their interests without some type of long-range strategic plan? Probably very few—if any. Common sense and necessity seem to dictate that in this time of increased competition for water we would be doing everything possible to regionalize and coordinate water management based on watershed boundaries and federal water quality standards.

Depending on the country involved, water problems and available solutions differ. This is especially the case between lesser-developed countries and the rest of the world. There is some irony here. For example, in the lesser-developed countries there may be enough water, but little money available to develop water storage and distribution systems. In more developed countries the problem is not money, the problem is availability.

To ensure adequate water supplies, the region involved must incorporate the following steps into its water resource management plan:

(1) Proper watershed management: this includes protecting aquifer recharge areas and upland vegetation so that rainfall can be held upstream and released gradually to meet the downstream needs.

(2) Conserve water and avoid needless waste: this is accomplished by pricing water, a valuable resource, to reflect the cost of supply.

In conserving water and avoiding needless waste, several steps can be taken. One step is to find a way to improve the efficiency of water use in agriculture. As pointed out earlier, irrigation for food production accounts for nearly three quarters of world water use and, worldwide, irrigation systems on average are only about 40% efficient. Elimination of excessive irrigation subsidies would greatly improve the efficiency of water use. New technological developments are helping in this effort. For example, micro-irrigation methods such as the *trickle-drip system* and precise scheduling of water deliveries can cut losses dramatically (WRI & IIED, 1988). Moreover, encouragement of the reuse of wastewater would greatly improve the efficiency of water use (more will be said about this important topic later).

Water use in the household can also be cut to aid in water conservation. Simple water-saving devices such as more efficient faucets, toilets, showerheads, and clothes washers can cut household use by one fifth. In addition, the use of water meters can cut domestic water consumption by as much as 45% (WRI & IIED, 1987).

CHAPTER 11

Water Pollution

I counted two-and-seventy stenches,
All well defined, and several stinks.
The river Rhine, it is well known,
Doth wash your city of Cologne;
But tell me, nymphs! what power divine
Shall henceforth wash the river Rhine? (S. T. Coleridge, 1828)

11.1 INTRODUCTION

IN an earlier chapter it was pointed out that pollution is relative and hard to define. The point being that water too polluted for swimming may not be too polluted for fishing or boating. Porteous (1992) defines pollution as any adverse alteration to the environment by a pollutant. Peavy et al. (1985) define *water pollution* "as the presence in water of impurities in such quantity and of such nature as to impair the use of the water for a stated purpose" (p. 14). For the purposes of this text, water pollution is defined as any physical or chemical change in surface water or groundwater that can adversely affect living organisms. All of these definitions have merit; more importantly, they speak to the problem that is dealt with in this chapter. What they all say is that water pollution is that condition (any condition) that makes water unsuitable for an intended use.

Right about now the reader may be thinking to him/herself: What's the big deal? Isn't pollution everywhere, isn't it inevitable?

Some people have this mind-set because they have little difficulty in personally identifying and defining water pollution. They see it every day. They see the debris bobbing along the course of their neighborhood river or stream. They see the oil slicks that slither, shimmer, and glimmer their way down the water course, rafting dead birds and fish along the way. They see those signs that warn, "DANGER—Water Unsuitable for Any Use." Then they sense the odor and taste problem. When they draw a glass of water from their kitchen tap and discover the water has a distinct odor and/or taste, the water, to them, is obviously polluted.

The point of all this is quite simple. We are able, in many cases, to determine what water pollution is. That is, we can do this when our senses allow us to do so. There is a problem, however. Our senses do not necessarily detect biological and chemical pollution—not until we become sick or worse.

The author has another concern about water pollution. Namely, when water is polluted, *we do not know what we do not know*—especially about the real consequences of water pollution. Yes, we can identify water pollution. The problem is we are not sure how water pollution affects us in the long run. Most short-term results are quite obvious. Again, however, it is the long-term effects (measured in decades, centuries, and longer) that we are unsure about. When you get right down to it, maybe it is because we do not know what we do not know about water pollution that allows us to be unconcerned. The point is: We need to be concerned.

Talking about being concerned, we must also become more informed. We must also get smart and use common sense and "good science" in making decisions related to water pollution and its mitigation. Water pollution is one thing, coming up with decisions based on cheap shots from the hip is another.

What does this mean? Good question. What this means is probably best explained through a real-world example.

11.1.1 REAL-WORLD EXAMPLE

Environmental policymakers in the Commonwealth of Virginia came up with what is called the *Lower James River Tributary Strategy* on the subject of nitrogen from the Lower James River and other tributaries contaminating the Lower Chesapeake Bay Region. Nitrogen is a nutrient. When in excess, nitrogen is a pollutant. Some "theorists" jumped on nitrogen as being the cause of a decrease in the oyster population in the Lower Chesapeake Bay Region. Oysters are important to the local region. They are important for economical and other reasons. From an environmental point of view, oysters are important to the Lower Chesapeake Bay Region because they have worked to maintain relatively clean Bay water in the past. Oysters are filter-feeders. They suck in water and its accompanying nutrients and other substances. The oyster sorts out the ingredients in the water and uses those nutrients it needs to sustain its life. Impurities (pollutants) are aggregated into a sort of ball that is excreted by the oyster back into the James River.

You must understand that there was a time, not all that long ago (maybe 45 years ago) when oysters thrived in the Lower Chesapeake Bay. Because they were so abundant, these filter-feeders were able to take in turbid Bay water and turn it almost clear in a matter of three days. (How could anyone dredge up, clean, and then eat such a wonderful natural vacuum cleaner?)

Of course, this is not the case today. The oysters are almost all gone. Where did they go? Who knows?

The point is that they are no longer thriving, no longer colonizing the Lower Chesapeake Bay Region in the numbers they did in the past. Thus, they are no longer providing economic stability to watermen; moreover, they are no longer cleaning the Bay.

Ah! But don't panic! The culprit is at hand; it has been identified. The "environmentalists" know the answer—they say it has to be nutrient contamination; namely, nitrogen is the culprit. Right?

Not so fast.

A Regional Sanitation Authority and a university in the Lower Chesapeake Bay region formed a study group to formally, professionally, and scientifically study this problem. Over a five-year period, using Biological Nutrient Removal (BNR) techniques at a local wastewater treatment facility, it was determined that the effluent leaving the treatment plant and entering the Lower James River consistently contained below 8 mg/L nitrogen for five consecutive years.

The first question is: Has the water in the Chesapeake Bay become cleaner, clearer?

The second question is: Have the oysters returned?

Answer to both questions, respectively: no; not really.

Wait a minute. The environmentalists, the regulators, and other well-meaning interlopers stated that the problem was nitrogen. If nitrogen levels have been reduced in the Lower James River, shouldn't the oysters start thriving, colonizing, and cleaning the Lower Chesapeake Bay again?

You might think so, but they are not. It is true that the nitrogen level in the wastewater effluent was significantly lowered through treatment. It is also true that a major point source contributor of nitrogen was reduced with a corresponding decrease in the nitrogen level in the Lower Chesapeake Bay.

If the nitrogen level has decreased, then where are the oysters?

A more important question is: What is the real problem?

The truth is that no one at this point and time can give a definitive answer to this question.

However, there are a number of questions that need to be answered: (1) Is nitrogen from the Lower James River and other tributaries feeding the Chesapeake Bay having an impact on the Bay (and the oysters)? (2) Is there evidence of low dissolved oxygen in the Lower James River? (3) Although there are high concentrations of nitrogen in the Lower James River, are there corresponding high levels of plankton (chlorophyll "a")? (4) Is it true that removing nitrogen for the sake of removing nitrogen would produce no environmental benefits, be very expensive, and divert valuable resources from other significant environmental issues?

Back to the problem with the decrease in oyster population in the Lower

James River/Chesapeake Bay region. Why has the oyster population decreased?

One theory states that because the tributaries feeding the Lower Chesapeake Bay (including the James River) carry megatons of sediments into the Bay, they are adding to the Bay's turbidity problem. When waters are highly turbid, oysters do the best they can to filter out the sediments but eventually they decrease in numbers and then fade into the abyss.

Is this the answer? That is, is the problem with the Lower Chesapeake Bay and its oyster population related to turbidity?

Only solid, legitimate, careful scientific analysis may provide the answer.

One thing is certain: Before we leap into decisions that are ill-advised, that are based on anything but sound science, and that "feel" good, we need to step back and size up the situation. This sizing-up procedure can be correctly accomplished only through the use of scientific methods.

Don't we already have too many dysfunctional managers making too many dysfunctional decisions that result in harebrained, dysfunctional analysis—and results?

Obviously, there is no question that we need to stop the pollution of our surface water bodies.

However, shouldn't we replace the timeworn and frustrating position that "we must start somewhere" with good common sense and legitimate science?

The bottom line: We shouldn't do anything to our environment until science supports the investment. Shouldn't we do it right?

11.2 MAJOR WATER POLLUTANTS

Water pollution may occur in many different forms and originate from many different sources. Oxygen-demanding wastes, disease-causing agents, inorganic chemicals and minerals, radioactive substances, thermal pollution, and organic chemical forms of water pollution are discussed in the following sections.

11.2.1 OXYGEN-DEMANDING WASTES

Sources of oxygen-demanding wastes include natural runoff from land, human sewage, animal wastes, decaying plant life, industrial wastes from oil refineries, food processing plants, and paper mills, and urban storm runoff. The problem with oxygen-demanding wastes is that they deplete dissolved oxygen in water causing fish to die or to migrate elsewhere. In addition, plant life is destroyed, foul odors generated, and livestock poisoned.

Control measures include minimizing or limiting agricultural runoff and wastewater treatment.

Water pollution from agricultural runoff includes pesticides, fertilizers, and animal wastes. There is no question about the need for pesticides; they are essential to food production and other crop products to produce the yield required to feed us. However, when pesticides are used in appreciable concentrations they can pose critical health hazards. Moreover, pesticides, although they are designed for a specific pest, often kill nonpest species such as birds and fish. Like pesticides, fertilizers are used to increase food production. Water pollution results from the nitrates and phosphates present in fertilizers. Nitrates, in sufficient concentrations, are toxic to animals and humans. Nitrates and phosphates contribute to the excessive growth of microscopic plant algae in some surface water bodies. Animal wastes are not only a source of water pollution but also create a potential health hazard.

11.2.2 DISEASE-CAUSING AGENTS

Disease-causing agents are a result of domestic sewage and animal wastes. They are responsible for causing outbreaks of waterborne diseases such as typhoid, cholera, infectious hepatitis, and dysentery. In addition, waterborne disease-causing agents infect livestock. Wastewater treatment, minimizing agricultural runoff, and establishing a dual water and waste disposal system are a few of the measures used to control disease-causing agents in water.

11.2.3 INORGANIC CHEMICALS AND MINERALS

Water pollutants in the form of inorganic chemicals and minerals include acids, salts, lead, mercury, plant nutrients, and sediments. Acids enter surface waters and groundwater from mine drainage, industrial wastes, and acid deposition. The problem with acid infiltration in water is its tendency to increase solubility of other harmful minerals. Acids also kill some organisms. In order to prevent acid pollution mines must be properly sealed, wastewater treatment should be employed, and atmospheric emissions of sulfur and nitrogen oxides must be reduced.

Salt contamination in water results from natural runoff from land, irrigation, mining, industrial wastes, oil fields, urban storm runoff, and from residues leftover from deicing roads. Salt kills freshwater organisms, causes salinity buildup in soil, and makes water unfit for domestic use, irrigation, and many industrial uses. In order to control salt contamination in water, wastewater must be treated, mined land should be reclaimed, and brine effluents should be banned from oil fields.

Problems with high lead content in surface waters is the result of lead smelting practices, pesticide use, and use of leaded gasoline. Lead is toxic to many organisms, including humans. To control lead contamination in water, wastewater treatment is required. In addition, the use of lead should be reduced to essential uses only.

As a form of water pollution, mercury compounds find their way into the food chains of wildlife, particularly fish. When these contaminated fish are consumed by humans they can be highly toxic. Mercury enters water systems via natural evaporation, industrial wastes, and fungicides.

Plant nutrients such as phosphates and nitrates were discussed earlier.

Sediments are the major source of water pollution. They enter water systems through natural erosion, poor soil management, and runoff from agricultural, mining, forestry, and construction activities. The problem with sediment contamination in water is that it fills in waterways, harbors, and reservoirs. Sediments also reduce fish and shellfish populations. They also work to reduce the ability of water to assimilate oxygen-demanding wastes. Obviously, the measure that should be employed to control sediment contamination in water is to employ more extensive soil conservation practices.

11.2.4 RADIOACTIVE SUBSTANCES

Radioactive substances are another type of pollutant that can affect aquatic ecosystems. Natural radioactivity occurs in the environment (rocks and soils); however, radiation pollution arises from the use of radioactive materials. Three main sources of radiation pollution are nuclear power plants, coal-fired power plants, and nuclear explosions. Radiation pollution can cause genetic defects and cancer. In order to control radiation pollution, the use of nuclear power plants must be strictly regulated and the processing and shipping of nuclear fuels and wastes must be strictly controlled.

11.2.5 THERMAL POLLUTION

Simply stated: *Thermal pollution* occurs when industry returns heated water to its source. Large-scale generation of electricity requires enormous quantities of water for cooling—water that is sometimes drawn from lakes, rivers, and streams. Power plants that burn fossil fuels or that use nuclear fuel generate great amounts of waste heat, some of which is removed by circulating cool water around and through hot power-generating equipment. Thus the heat is transferred to the water, raising its temperature. When the heated water is discharged, it can have an adverse effect on aquatic ecosystems. For example, thermally polluted water can decrease solubility of oxygen in water, can kill some fish, can increase the susceptibility of some aquatic organisms to parasites, disease, and chemical toxins. Moreover, thermal pollution, generally, changes composition of and disrupts aquatic ecosystems.

Cooling water used by industry does not have to be released into aquatic ecosystems. There are other methods of discharging the heat. One method is to return heated water to ponds or canals. A second method is to transfer waste heat to the air or to use it to heat homes and other buildings.

11.2.6 ORGANIC CHEMICALS

Certain organic chemicals are particularly dangerous threats to aquatic environments because they bioaccumulate through food webs. The chemicals in this category that cause the most concern include oil and grease, pesticides and herbicides, plastics, detergents (phosphates), and chlorine compounds.

11.2.6.1 Oil and Grease

Oil and grease enter aquatic ecosystems via machine and automobile wastes, pipeline breaks, offshore oil well blowouts, natural ocean seepages, tanker spills, and cleaning operations.

Along with causing taste and odor problems, oil and/or grease pollutants can cause disruption of aquatic ecosystems. They also cause economic, recreational, and aesthetic damage to coasts, fish, and waterfowl. Simply stated: Aquatic life is stifled by oil and grease, which coat practically everything with which they come into contact, including wildlife.

Oil pollution is an ever-present threat to surface waters. This is especially the case in the river waterways that are used to transport it. During the past several years, several major oil spills have occurred and made headlines worldwide. In order to prevent these catastrophic events from occurring strict regulation of oil drilling, transportation, and storage must be effected. In addition, it is important to institute control measures such as requiring the collection and reprocessing of oil and greases from service stations and industry. Another control measure that has been gaining momentum and attention is oil spill mitigation. Continued research is needed on developing additional less costly and more efficient means to contain and mop up spills. "Contain" is the key word here. When oil tanks and other vessels are constructed, some type of containment system should be "built in" as part of the construction project. However, this is easier said than done in some applications. For instance, it is difficult to build containment around a cargo tanker. Transportation of oil poses an environmental problem, especially by ship or tanker. Double hulls and segregated tank arrangements have been employed in this effort; however, with the exception of oil spill booms, other types of mechanical containment systems for tanker spills are difficult if not impossible to employ.

11.2.6.2 Pesticides and Herbicides

Earlier it was pointed out that certain organic chemicals are dangerous threats to the aquatic environment because they bioaccumulate through food webs. Pesticides and herbicides are chemicals in this category. In the United States, for example, enormous quantities of pesticides are used by farmers, foresters, and mosquito control agents to control weeds, insects, rodents, and disease-producing fungi. These pesticides and herbicides enter rivers and streams through runoff, through the effluents of pesticide/herbicide manufacturing plants, from drifting spray mists during application, through the washing and maintenance of spraying equipment, and by accidental discharges. Moreover, when soil sediments are contaminated with pesticides/herbicides and then are washed off into streams, stream contamination is likely.

Many pesticides and herbicides are toxic or harmful to fish, shellfish, predatory birds, and mammals. These agents tend to concentrate in human fat cells and can be toxic; they can cause birth and genetic defects and cancer.

In order to protect surface waters from contamination with pesticides and herbicides several steps should be taken. The most obvious and useful step is to ban and/or reduce use of those pesticides and herbicides that are the most dangerous. Users should be encouraged or required to switch to alternative methods of insect control (biological and ecological methods). Practices that include modifying cultivation procedures, biological control, genetic control by sterilization, genetic control by breeding resistant crops and animals, chemical control using natural sex attractants and hormones, irradiation foods, integrated pest management, and changing the attitudes of consumers and farmers are being attempted in an effort to reduce the use of toxic pesticides and herbicides.

11.2.6.3 Plastics

The magnitude of increased plastics production and use in recent years has created a significant waste disposal problem. The problem with plastics, from an environmental and aquatic lifeform point of view, is that most of them are not biodegradable. Even if they do biodegrade the process is very slow. Plastics enter surface waters either from industrial or domestic sources. Plastics discarded in surface waters may remain in largely unaltered form for years if not decades (Laws, 1993).

The effects of plastics pollution in surface waters range from those of

aesthetics to fish kills. It is not uncommon today to visit just about any surface water body and find plastics either bobbing in the water or washed up
against the stream or river bank. Whenever these plastic discards are so obvious and apparent to on-lookers, it brings to mind the old practice (saying): "I don't want or need it any longer—take it to the river and dump it"
(Spellman, 1996b).

Plastics kill fish. Fish are killed by eating plastic; they are also killed
when they become ensnared with plastic debris. When fish ingest plastic
debris, no one really knows how many actually die as a result; furthermore,
we will probably never know—it would take an autopsy of each dead
aquatic organism to confirm the problem. Have no doubt, however; plastic
debris kills fish.

In order to control the plastic debris problem in surface waters it is
necessary to ban dumping. Another useful step is to encourage recycling of plastics. Moreover, reducing the use of plastics in packaging
would help.

11.2.6.4 Detergents (Phosphates)

In a general sense, a *detergent* is a cleansing agent. Soap is a cleansing
agent and detergent. Soap is made by combining either oils or fats with sodium or potassium hydroxide (alkalies). Soap molecules in water effec-
tively interact with oil and grease so that both are carried away with dirt in
this solution. This process is the *detergent action* of soap.

Ordinary soap causes few problems in wastewater treatment; it is normally degraded into harmless substances by bacteria in sewage. This is not
the case, however, in water with dissolved mineral content; minerals impair the detergent action of soap. To alleviate this problem, synthetic detergents were developed. These synthetic detergents react with the dissolved
minerals without greatly impairing the cleansing action. Unfortunately, it
wasn't long before it became common knowledge that synthetic detergents
are nonbiodegradable and difficult to remove by treatment processes.

Because they are difficult to remove from the wastestream, synthetic detergents persist when discharged into receiving waters. These "hard" detergents cause mountains of foam and suds to be formed in the receiving water
body. Fortunately, the use of hard detergents has been discontinued and replaced by "soft" biodegradable detergents. There is still some concern,
however, about possible yet unknown (the we don't know what we don't
know syndrome) consequences of the decomposition products on the environment.

Another pollution problem results whenever phosphate detergents are
used for dishwashing and laundering. Even though it is true that phosphate
detergents make these activities more efficient, it is also true that they are

responsible for as much as 50% of the phosphorus in municipal wastewaters.

In order to gain understanding of why some detergents still contain appreciable amounts of phosphorous it is important to point out why phosphates are in detergents. Typically, a detergent contains two basic components, a *surfactant* and a *detergent builder*, which together often account for most of the remaining weight of the detergent (Layman, 1984). Surfactants (surface-active agents) are soluble compounds that reduce the surface tension of liquids, or reduce interfacial tension between two liquids or a liquid and a solid. Surfactants consist primarily of bipolar components such as fatty alcohol ethoxylates, fatty alcohol sulfates, and alkyl-sulfonates. Their bipolar characteristic allows them to penetrate between dirt and fabric. The way this typically works is that one side of the surfactant molecule is attracted to the fabric (or material being cleaned); the other side is attracted to the contaminant (dirt). Because of their ability to lower surface tension, surfactants aid in lifting off grease. The detergent builder works to soften the water by removing calcium and magnesium ions, preventing them from combining. This prevention mechanism is important, otherwise the ions would combine with the surfactant to form curd (gummy precipitate) that is not easily removed with rinsing water. Moreover, the gummy precipitate settles in clothes, and forms a film that may build up in the washer, ultimately clogging the machine.

Phosphate compounds such as sodium tripolyphosphates are commonly used as detergent builders. Sodium tripolyphosphate forms compounds with the calcium and magnesium ions in hard water. This prevents the ions from reacting with the soap and reducing its detergency. As you might expect, wastewater ends up with a high phosphate content. The bad news is that the phosphates are not easily removed by ordinary waste treatment processes.

It is interesting to note, as Laws (1993) points out, that if sodium tripolyphosphate or some other detergent builder were removed from detergents without substituting another builder, it would still be possible to get a clean wash under any of the following conditions:

(1) About 10 times as much detergent was added to the washing machine (Hammond, 1971).
(2) A water softening agent such as sodium carbonate was added to the washing machine before both the wash and rinse cycles.
(3) A water softener had been used to remove the calcium and magnesium from the water before use of the water in the washing machine.
(4) The washload contained a naturally low concentration of magnesium and calcium. (p. 149)

Although various substitute compounds have been tried in place of the detergent builder sodium tripolyphosphate and some areas have attempted to ban the use of detergent builders altogether, the fact of the matter is unless consumers can wash clothes in any of the above described situations, they will not be able to wash clothes to the "whiter-than-white" that they demand.

At this point the reader, hopefully, has some understanding as to why phosphate detergent builders are still being utilized.

11.3 SOURCES OF WATER POLLUTION

Sources of water pollution are distinguished between *point sources* and *nonpoint sources;* they can be man-made or natural. A *point source* is a source that discharges pollutants, or any effluent, such as wastewater through pipes, sewers, and channels into bodies of water at specific locations. Examples include sewage treatment plants, factories, utilities, offshore oil wells, oil tankers, and underground coal mines. Thermal pollution is an example of point source pollution. Thermal pollution often occurs near chemical or nuclear power plants. The impact of thermal pollution includes killing of plants and a reduction in the amount of oxygen needed to sustain aquatic life. Industrial and municipal point sources are easy to identify; thus, most water pollution control efforts have concentrated on these entities.

A *nonpoint source* of water pollution is usually easier to spot (a stream is cloudy or discolored) than it is to control. Simply stated: Nonpoint pollution is a source that is one of many widely scattered (diverse) sources that discharge pollutants over large areas. Examples include runoff into surface water and seepage into groundwater from logged forests, livestock feedlots, croplands, urban and suburban lands, construction areas, parking lots, and roadways.

According to the WEF (1992), nonpoint source water pollution pollutes 50% of the swimming water in the United States. Moreover, it contaminates groundwater that makes up 95% of the country's freshwater supply. In addition, it adds the salts, pesticides, and toxic chemicals that contaminate fish and shellfish, causing major economic loss plus risks to human health.

11.4 WATER QUALITY MANAGEMENT

Simply stated: *water quality management* is concerned with the control of pollution from human activity so that the water is not degraded to the point where it is no longer suitable for intended uses (Davis & Cornwell, 1991). What this really means is that every effort must be made to safeguard water quality. To ensure acceptable water quality for human consumption and ecosystem survival, the highest priority should go to programs that reduce the generation of wastes from industrial plants, electric power generation, cities and towns, mining and smelting operations, and from agriculture. Moreover, for those wastes that cannot be eliminated there must be adequate containment and treatment.

Containment includes agriculture where more efficient irrigation methods, control of runoff from animal feed lots, and the exercise of much greater caution in the use of chemical fertilizers, pesticides, and herbicides should be utilized.

In treatment of municipal sewage, innovative, cost-effective approaches must be developed and utilized. This is especially the case for Third World countries that are beginning to deal with sewage problems.

If valuable water resources and fragile ecosystems worldwide are to be protected, all countries must chip in to restore degraded watersheds, lakes, rivers and streams, and wetlands. Moreover, appropriate waste reduction and pollution prevention measures must be instituted.

In the United States, it was not until the early 1970s that the public exerted enough pressure on Congress to pass the Safe Drinking Water Act (SDWA). This Act required USEPA to nationalize standards, called maximum contaminant levels (MCL), for the quality and treatment of water supplies, regulate underground waste injections, and set monitoring and reporting regulations for public water systems (Renew America, 1989). Since passage of the SDWA, other important regulations have been put in place such as the Clean Water Act (CWA). Since 1972 several improvements in water quality have been seen. It is important to point out, however, that all the regulations in the world will not have as much impact on water conservation as public acceptance and compliance. The fact is an estimated 40+% of the water used in the United States is unnecessarily wasted. The major reason for the large amount of water wasted in the United States is that this same government that regulates water quality also keeps water prices artificially low, in an attempt to stimulate economic growth and by using taxes to build dams and water transfer projects. Thus, the taxpayers are paying a heavy price for water, but they do not realize it.

The bottom line: Until the public experiences water shortages, decreased water quality, and/or thirst, water conservation is not likely to receive the priority status it deserves. Earlier it was pointed out that water is special, strange, and different. We should also add that water is critical to our very survival.

When will we recognize our inextricable dependence on water?

Another important question is: Will we wait until drought turns our

Another important question is: Will we wait until drought turns our countryside into a dust bowl of ruined industry, economic havoc, and shattered lives?

Then the ultimate question becomes: Will we wait until it is too late to act—until water is beyond being strange, different, and special—to the point when it is rare, very rare, to the point of being no longer available for use?

I am not sure when or if we will answer these questions. I only know that if these questions are not answered soon, then it may become too late to answer any of them in a fashion that will ensure sustainment of life as we know it.

11.5 THE RHINE RIVER DISASTER

This chapter opened with an 1828 description of the Rhine River by S. T. Coleridge. The question is: Has the Rhine River improved any since Coleridge's time? If the key word is "any," then the answer is yes—sort of.

Let's take a look at the Rhine River, past and present.

The Rhine River winds 820 miles (1,320 kilometers) through Switzerland, France, Germany, the Netherlands and empties into the North Sea. Numerous cities and industrial complexes (chemical, steel, etc.) border this heavily used and abused river.

In early 1970 the Rhine was so heavily polluted with oxygen-demanding wastes and toxins that it was devoid of most aquatic life. Cleanup efforts were begun in 1970 and soon DO levels increased, BOD decreased, and the number of types of aquatic life increased significantly.

Unfortunately, in 1986 all this progress was set back by a fire and subsequent chemical spill (disaster) near Basel, Switzerland.

Because of the spill, highly hazardous chemicals entered the Rhine and flowed into the already polluted North Sea. Cities in the four countries depending on the river for drinking water had to temporarily find other supplies. Livestock that drank water from the river died. Fishing was banned. More than half a million aquatic animals died. Several miles of river suffered severe ecological damage.

There is another problem with this spill; that is, not all the damage has been felt or realized yet. For example, several pounds of mercury were spilled into the river, most of which settled to the bottom. Scientists fear that much of this mercury will be biologically amplified in the food webs. At this point in time we are not certain as to the entire impact that will be felt because of this disaster—again, we do not know what we do not know.

One thing is certain. Coleridge knew something in 1828 that is relevant today: "What power divine shall henceforth wash the river Rhine?"

CHAPTER 12

Water Reuse

Planned reuse of municipal wastewater has been practiced throughout the world for many years. Interest in reuse is increasing as existing water supplies are unable to meet the demands of urban growth. Both the water and wastewater industries recognize the value of water reclamation and reuse as an integral part of water resources management. (WERF, 1994, p. ES-1)

12.1 INTRODUCTION

WHEN viewed from a holistic point of view (i.e., worldwide), water is a limited resource. The average person might ask: How can this be? We are literally shrouded in water. Water covers most of the earth—water, water, water, everywhere you look there is water. Obviously, this person does not live in or is not familiar with arid and semi-arid parts of the globe. Maybe our viewer is referring to the hydrologic cycle—that natural process of rainfall-runoff-evaporation, which repeats itself continuously (we can only hope that it continues to do so). The point is that our viewer is not alone in his/her assessment of water—the state of water. The fact is most people do not give water a second thought. Most of us don't give it any thought at all.

Not yet we don't.

Why not? Because we just don't understand the problem. Just how serious is this so-called water problem, shortage? Very serious.

For example, consider the following article that appeared in *The Virginian-Pilot* on January 24, 1997.

WATER SHORTAGE FORECAST BY 2025 FOR MUCH OF WORLD

"About two-thirds of the world's population will suffer from water shortages by the year 2025 unless action is taken to stem the wasting and polluting of water, a U.N.-commissioned report says.

The report . . . was done by the Stockholm Environment Institute for the United Nations Commission on Sustainable Development.

221

*The report notes that water use has been growing at more than twice the rate
of population increase in the 20th century and predicts that about 2.5 billion
people, or about two of every three people, will live in water-poor areas
within the next 30 years.*

*The report found that 20 percent of the world's people in 1995 did not have
access to safe drinking water and that half lacked proper sanitation. (p.
A-6)*

To awaken the masses to the dilemma that lies in waiting out there ahead
of us, it will take an event—maybe it will be an event as dramatic and
sweeping as that fungus that swept Ireland in 1850—and like the blight,
drought will be the enormous wind behind the human back driving action,
action to preserve what we cannot afford not to preserve: *WATER.*

12.2 WATER REUSE/RECLAMATION

Water is limited. Maybe it would be more correct to say that in many locations throughout the world, water is limited. Today, many United States
water supplies, like many others worldwide, are threatened by drought.
The point is that without adequate water, parks, nurseries, industry, agriculture, public landscapes, and golf courses will suffer dramatic losses.

In order to reduce these dramatic losses, prudent use of water resources
must be effected. More specifically, we must make water reuse an increasingly common component of water resource planning. This is especially
the case in light of the present and projected future trend pointing to a decrease in the opportunity to develop conventional water supplies and the
increasing costs associated with wastewater treatment (Hamann &
McEwen, 1991).

When we talk about water or wastewater reuse, what are we really talking about? Well, in the first place, wastewater reuse is nothing new. Discharging wastewater to water courses has been practiced for several years.
These same water courses, downstream, served as sources of water supply
for several communities. The problem is that with the growth in urban
population there has been an accompanying growth in urban sewerage discharge of treated and untreated wastewaters to these same water courses.
Thus, for the users downstream what this really meant was that their mor-
tality rates increased due to cholera, typhoid, and other diarrheal diseases
(Okun, 1996). Fortunately, the considerable health impact of the polluted
water courses was mitigated around 1900 through the development of filtration and chlorination that sharply reduced enteric disease in most indus-
trialized nations. This is not the case, however, in other parts of the world
where it is well documented that waterborne disease rates continue to be
high and the threats of epidemics are ever present when such contaminated

waters are used as sources of water for drinking and for irrigation of market crops (Okun, 1996).

Because population growth continues to soar, with its accompanying increase in contamination of both surface and groundwater supplies, along with uneven distribution of water resources and with periodic droughts, water reclamation and reuse are important.

We have been forced to search for innovative sources of water. Today, in water and wastewater management the key words are: *beneficial reuse*. The fact of the matter is we have at our "disposal" a valuable reuse product: highly treated wastewater effluent. However, it should be pointed out that nothing, absolutely nothing, should substitute for efficient use of existing water supplies, water conservation, and development of new water resources. We should take care of what we have. We should appreciate water for what it really is—that special liquid substance that we cannot live without. Are we to let water go the route of gasoline? That is, can you imagine the day when a gallon of water will cost more than a gallon of gasoline?

12.2.1 Reuse Terms

When we refer to water reuse and reclamation we are referring to reclaimed water applications that include urban reuse, industrial reuse, agricultural irrigation, groundwater recharge, potable water supply augmentation, habitat restoration, and other miscellaneous uses. Every branch of technology and science has its own language for communication, and water reclamation and reuse is no different. In order to understand water reclamation and reuse, it is necessary for the reader to acquire a slight vocabulary as to the important concepts related to the topic. Thus, there are several terms that are frequently used in the field of wastewater reclamation and reuse that are important. These terms are from Metcalf & Eddy (1991, pp. 1139–1140) and are discussed (defined) in the following:

- *Beneficial uses*   the many ways water can be used, either directly by people or for their overall benefit.
- *Direct potable reuse*   the piped connection of water recovered from wastewater to a potable water-supply distribution system or a water treatment plant, often implying the blending of reclaimed wastewater.
- *Direct reuse*   the use of reclaimed wastewater that has been transported from a wastewater reclamation plant to the water reuse site without intervening discharge to a natural body of water (e.g., agricultural and landscape irrigation).
- *Indirect potable reuse*   the potable reuse by incorporation of re-

claimed wastewater into a raw water supply; the wastewater effluent is discharged to the water source, mixed and assimilated with it,
with the intent of reusing the water instead of as a means of disposal. This type of potable reuse is becoming more common as water resources become less plentiful.

- *Indirect reuse* — the use of wastewater reclaimed indirectly by passing it through a natural body of water or use of groundwater that has
  been recharged with reclaimed wastewater. This type of potable reuse commonly occurs whenever an upstream water user discharges
  wastewater effluent into a water course that serves as a water supply
  for a downstream user.
- *Planned reuse* — the deliberate direct or indirect use of reclaimed
  wastewater without relinquishing control over the water during its
  delivery.
- *Potable water reuse* — a direct or indirect augmentation of drinking
  water with reclaimed wastewater that is normally highly treated to
  protect public health.
- *Reclaimed wastewater* — wastewater that, as a result of wastewater
  reclamation, is suitable for a direct beneficial use or a controlled use
  that would not otherwise occur.
- *Wastewater reclamation* — the treatment or processing of wastewater to make it reusable.
- *Wastewater reuse* — the use of treated wastewater for a beneficial
  use such as industrial cooling.

12.3 REUSE FOR NONPOTABLE PURPOSES

Today, reclamation of wastewater and its reuse for nonpotable purposes
in cities is normally accomplished through use of dual distribution systems.
According to the WERF (1994), *dual distribution systems* "distribute two
grades of water—one potable and the other nonpotable—to the same service area. The quality, quantity, and pressure available in each are functions
of the sources and intended uses for each grade of water" (pp. 2–5). The
driving force behind the trend toward reuse of nonpotable water is a combination of two important factors: (1) it provides an additional source of water and (2) it reduces the cost of wastewater disposal (Okun, 1996).

The rapid increase in population growth, especially in urban areas, has
resulted in inadequate sources of quality waters. Moreover, even those areas that have followed public health principles and developed protected upland sources are finding that their sources no longer provide adequate supplies. There are several reasons these areas have had difficulty in adding to
their supplies: (1) additional sources are more distant, (2) they are more

costly to develop, and (3) it has become increasingly more difficult to obtain water rights to additional sources.

Obviously, this situation (i.e., the real need for additional water) calls for alternatives. Fortunately, there is, at present, a movement toward recognizing this need and the need for alternatives. The mindset is shifting to one that professes that no higher quality water, unless there is a surplus of it, should be used for a purpose that can tolerate a lower grade. This policy makes sense. This is especially the case when one considers that only a small amount of the water used in urban settings is required for potable use. When reclaimed water is used in urban settings it allows for the conservation of limited high quality waters that can be used to serve a much larger population. This is where the development of dual distribution systems comes into play. Heavily populated areas are beginning to construct dual distribution systems: One system, from a high quality source, for potable purposes and the other, reclaimed water also of high quality but not suitable for drinking, for nonpotable purposes in households, industry, commerce, and public facilities, including landscape irrigation, lakes, public fountains, and environmental improvements.

The need for additional public water supplies has emerged as the primary factor (i.e., when the public wants water, they demand it—they have a right to it) in driving the push for investing in water reclamation and reuse projects in urban areas. However, it should also be pointed out that reducing the cost of wastewater treatment and disposal is another important factor. This makes sense when you consider the overall expense that is added to the treatment and disposal of wastewater by regulations. In particular, regulations have been enacted, for example, that require nutrient removal. It is less expensive to reclaim wastewater for nonpotable reuse than it is to treat it for discharge.

USEPA's *Guidelines for Water Reuse* (1992) fully covers the current status of practice of urban nonpotable reuse. This 247-page publication includes all types of reclamation and reuse and lists regulations and standards, state by state where they exist, as well as covering international practices.

12.3.1 NONPOTABLE URBAN USES

Metcalf & Eddy (1991) point out that "irrigation of landscaped areas and golf courses in the urban environment has become an important use of reclaimed wastewater in recent years" (p. 1144). Along with irrigation of landscaped areas such as parks, athletic fields, school yards, areas around public facilities and buildings, and highway medians and shoulders, there are other urban uses of reclaimed water. For example, irrigation of landscaped areas around family homes and nurseries can be included. Moreover, reclaimed water can be used for commercial activities such as vehicle washing, window washing, and concrete production. Reclaimed wastewater can be used in the fire protection service. This is an important use because it not only allows for a dedicated fire main that separates it from potable water mains but also it conserves a large volume of potable water. Reclaimed water can also be used for toilet and urinal-flushing in industrial, commercial, and residential buildings, especially in those with occupancies (e.g., multistory facilities). Reclaimed wastewater can also be used in industries for such applications as boiler-feed water, make-up water for evaporative cooling towers, and irrigation of facility grounds. Specifically, we are talking about industries such as textiles, chemical manufacture, petroleum products, steel manufacture and other industrial applications where the product would pose no health threat to consumers.

12.3.2 WATER QUALITY REQUIREMENTS OF RECLAIMED WATER FOR URBAN NONPOTABLE REUSE

Regulations for reclaimed water used for nonpotable purposes have been adopted by several states. However, to date, there are no federal regulations for nonpotable reuse. It should be pointed out, however, that USEPA (1992) specified degrees of treatment and quality standards for a wide range of uses with the most stringent being for the unrestricted use where large populations are likely to be exposed and for irrigation of so-called *market crops*. In brief, the USEPA requires that treatment include the filtration of secondary effluents followed by disinfection to produce water with average turbidity equal to or less than 2 NTU, and a maximum not to exceed 5 NTU, with no detectable fecal coliform in 100 ml, and a minimum chlorine residual of 1 mg/L after a minimum contact time of 30 minutes.

If water quality requirements can be maintained for reclaimed wastewater that is designated for reuse, and if population centers continue to grow at record rates, then it logically follows that there will be a corresponding growth of nonpotable reuse in the future.

12.4 POTABLE WATER REUSE

The need for additional water supplies, especially in the arid regions of the western and southwestern United States, continues to increase and the pressure to develop alternate supplies is constant. The issue which drives the technology and the public health science is the conflict between the traditional approach and to obtain drinking water supplies from the "best available source," as suggested by the US Public Health Service and the US Environmental Protection Agency, and the desire to develop new supplies from reclaimed water. (Pla, Grebbien, & Gaston, 1995, p. 715)

According to Rowe & Abdel-Magid (1995), by the year 2000, in the United States, there will be a reduction in freshwater withdrawals. This will occur primarily because of wastewater reclamation and reuse. The question is: How much of this reclaimed water will be reused for potable water? That is, how much of this reclaimed water will be discharged into a water course, lake, water supply reservoir, or underground and withdrawn downstream or downgradient at a later time for potable purposes? How much will be piped directly from plant effluent into the potable water system? This is hard to judge; we can only address potable water reuse at present.

Talking about the present, the attitude towards using reclaimed wastewater for potable water is cautious. There are two reasons for this attitude: (1) public acceptance and (2) health and safety concerns.

In regards to public acceptance, the problem is aesthetic concerns. It does not matter that a potable water reuse program can be technically viable, the recovered water proven safe, and regulatory agencies approving its use, a program can still fail because of lack of public acceptance. It is difficult for the public to accept the reuse of a product in which waste has been deposited. Obviously, what is required to overcome the public's skepticism, fear, and doubt is a public education program.

A public education program must address those issues of major concern to the public. This means informing the public (1) about the ability to recover potable from wastewater; (2) that the recovered water is safe for either direct or indirect human consumption; (3) that fail-safe, redundant operational procedures are available to assure product safety; (4) that water recovery treatment processes can be easily modified in response to new regulatory requirements; (5) that the distribution of recovered potable water will be equitable; and (6) that the cost and reduced environmental impact of potable water recovery compares favorably to other alternative water supplies.

According to the WERF (1994), "direct potable reuse is not practiced anywhere in the world" (pp. 3–71). Beyond the obvious aesthetic reasons that prohibit direct potable reuse, there is a more important aspect to be considered: the safety and health of the user. Considerable research is underway in an attempt to determine if this concern is viable. Current drinking water standards presume that water supplies are derived from relatively unpolluted freshwater sources. There is little doubt that great advances have been made in analytical methods for identifying contaminants in water, however, only a small fraction of the contaminants present in recycled water can be identified. Thus, the major constraint against pipe-to-pipe water supply reusing reclaimed wastewater for human consumption is the concern about chemical and pathogen transmission.

Making the ultimate decision to reuse treated wastewater in pipe-to-pipe, direct potable use comes down to one basic, limiting factor: We do not know what we do not know about the long-term consequences of human consumption of treated wastewater effluent. The point is we don't know at this moment in time, but we need to find out—soon!

References

Abrahamson, D. E. (ed.) (1988). *The Challenge of Global Warming*. Washington, DC: Island Press.

Allan, J. D. (1995). *Stream Ecology: Structure and Function of Running Waters.* London: Chapman & Hall.

Arasmith, S. (1993). *Introduction to Small Water Systems.* Albany, Oregon: ACR Publications.

Asano, T., Smith, R. G., & Tchobanglous, G. (1985). Municipal Wastewater: Treatment and Reclaimed Water Characteristics, *Irrigation with Reclaimed Municipal Wastewater: A Guidance Manual,* Pettygrove, G.S. & Asano, T. (eds.), Chelsea, MI: Lewis Publishers.

Asimov, I. (1989). *How Did We Find Out About Photosynthesis?* New York: Walker & Company.

ASTM (1969). *Manual on Water.* Philadelphia: American Society for Testing and Materials.

AWWA (1995a). *Basic Science Concepts and Applications* (2nd ed.). Denver: American Water Works Association.

AWWA (1995b). *Water Treatment* (2nd ed.). Denver: American Water Works Association.

Bartels, J. H. M., Burlingame, G. A., & Suffet, I. H. (1985). Flavor Profile Analysis: Taste and Odor Control of the Future, *Journal of American Water Works Association, 78,* 3, 50.

Benefield, L. D. & Randall, C. W. (1980). *Biological Process Design for Wastewater Treatment,* Englewood Cliffs, NJ: Prentice-Hall.

Benke, A. C. (1990). A Perspective on America's Vanishing Streams. J. *Am. Benthol. Soc., 9,* 77–88.

Bitton, G. (1994). *Wastewater Microbiology.* New York: John Wiley & Sons.

Bradbury, I. (1991). *The Biosphere.* New York: Belhaven Press.

Carson, R. (1962). *Silent Spring.* Boston: Houghton Mifflin Company.

Chen, K. Y., Young, C. S., Jan, T. K., & Rohatgi, N. (1974). Trace Metals in Wastewater Effluent, *J. Water Pollution Control Federation, 45,* 2663.

Clark, N. A., Stevenson, R. E., Chang, S. L., & Kobler, P. W. (1961). Removal of Enteric Viruses from Sewage by Activated Sludge Treatment, *J. Am. Public Health, 51,* (8), 1118.

Clary, D. (1997). What Makes Water Wet, *Geraghty & Miller Water Newsletter, 39,* 4.

Coackley, P. (1975). Developments in Our Knowledge of Sludge Dewatering Behavior, *8th Public Health Engineering Conference* Held in the Department of Civil Engineering, Loughborough: University of Technology.

Davis, M. L. & Cornwell, D. A. (1991). *Introduction to Environmental Engineering,* New York: McGraw-Hill, Inc.

Eckenfelder, W. W., (1989). *Industrial Water Pollution Control, Series in Water Resources and Environmental Engineering,* New York: McGraw-Hill.

Federal Register (1988). Secondary Treatment Regulations, 40 CFR Part 133, July.

Fortner, B. & Schechter, D. (1996). U.S. Water Quality Shows Little Improvement Over 1992 Inventory, *Water Environment & Technology, 8,* 2.

Geraghty & Miller (1997). *Water Newsletter, 39,* 4.

Gilcreas, F. W., Sanderson, W. W. & Elmer, R. P. (1953). Two New Methods for the Determination of Grease in Sewage, *Sewage Ind. Wastes, 25,* 1379.

Hamann, C. L. & McEwen, B. (1991). Potable Water Reuse, *Water Environment & Technology,* January.

Hammer, M. J. (1986). *Water and Wastewater Technology,* 2nd ed. New York: John Wiley & Sons.

Hammond, L. (1971). Phosphate Replacements: Problems with the Washday Miracle. *Science, 172,* 361–363.

Hickman, C. P., Roberts, L. S., & Hickman, F. M. (1988). *Integrated Principles of Zoology.* St Louis: Times Mirror/Mosby College Publishing.

Jones, F. E. (1992). *Evaporation of Water.* Chelsea, Michigan: Lewis Publishers, Inc.

Jost, N. J. (1992). Surface and Ground Water Pollution Control Technology. *Fundamentals of Environmental Science and Technology,* Porter-C. Knowles, Rockville, Maryland: Government Institutes, Inc.

Kemmer, F. N. (1979). *Water: The Universal Solvent,* 2nd ed. Oak Brook, Illinois: Nalco Chemical Company.

Kordon, C. (1993). *The Language of the Cell.* New York: McGraw-Hill, Inc.

Koren, H. (1991). *Handbook of Environmental Health and Safety: Principles and Practices.* Chelsea, Michigan: Lewis Publishers.

Laws, E. A. (1993). *Aquatic Pollution: An Introductory Text.* New York: John Wiley & Sons, Inc.

Layman, P. L. (1984). Brisk Detergency Activity Changes Picture for Chemical Suppliers. *Chem. and Eng. News,* Jan. 23, pp 17–20, 31–49.

Manual (1992). *Guidelines for Water Reuse.* EPA/625/R-92/004, Office of Water, U.S. Environmental Protection Agency, Washington, DC.

Masters, G. M. (1991). *Introduction to Environmental Engineering and Science.* Englewood Cliffs, NJ: Prentice-Hall.

McGhee, T. J. (1991). *Water Supply and Sewerage.* New York: McGraw-Hill, Inc.

Miller, G. T. (1988). *Environmental Science: An Introduction.* Belmont, California: Wadsworth Publishing Company.

Micklin, P. P. (1988). Desiccation of the Aral Sea: A Water Disaster in the Soviet Union. *Science, 241,* 1170–1176.

Metcalf & Eddy, Inc. (1991). *Wastewater Engineering: Treatment, Disposal, Reuse.* 3rd ed. New York: McGraw-Hill, Inc.

Moran, J. M., Morgan, M. D. & Wiersma, J. H. (1986). *Introduction to Environmental Science.* New York: W. H. Freeman and Company.

Myers, N. (1984) (ed.) *Gaia: An Atlas of Planet Management.* Garden City, N.Y.: Anchor Books.

National Academy of Science (1962). *National Research Council Publication.* 100-B, 1962.

Odum, E. P. (1971). *Fundamentals of Ecology.* Philadelphia: Saunders College Publishing.

Odum, E. P. (1983). *Basic Ecology.* Philadelphia: Saunders College Publishing.

Okun, D. A. (1996). A History of Nonpotable Urban Water Reuse. A paper presented at *Water Reuse Conference Proceeding,* San Diego, California, February 25–28.

Peavy, H. S., Rowe, D. R. & Tchobanglous, G. (1985). *Environmental Engineering.* New York: McGraw-Hill, Inc.

Pimentel, D. (1989). Waste in Agriculture and Food Sectors. Unpublished paper, Cornell University, College of Agriculture and Life Sciences.

Pla, M. M., Grebbien, V. & Gaston, J. M. (1996). Potable Reuse and the Emerging Conflicts with Drinking Water Regulations. A paper presented at Water Reuse Conference Proceedings, San Diego, California, February 25–28.

Porteous, A. (1992). *Dictionary of Environmental Science and Technology.* (Revised ed.). New York: John Wiley & Sons.

Postel, S. (1984). *Water: Rethinking Management in an Age of Scarcity.* Worldwatch Paper 62. Washington, DC: Worldwatch Institute.

Postel, S. (1985). Managing Freshwater Supplies, in Brown et al., *State of the World 1985.* New York: Norton.

Price, J. K. (1991). *Basic Math Concepts for Water and Wastewater Plant Operators.* Lancaster, PA: Technomic Publishing Company, Inc.

Price, P. W. (1984). *Insect Ecology.* New York: John Wiley & Sons, Inc.

Quagliano, J. V. (1964). *Chemistry.* 2nd ed. Englewood Cliffs, NJ: Prentice-Hall.

Quigg, P. W. (1976). *Water: The Essential Resource.* New York: National Audubon Society.

Renew America (1989). *The State of the States 1989.* Washington, D.C.

Rowe, D. R. & Abdel-Magid, I. M. (1995). *Handbook of Wastewater Reclamation and Reuse.* Boca Raton, Florida: Lewis Publishers.

Salvato, J. A. (1982). *Environmental Engineering and Sanitation.* 3rd ed. New York: John Wiley & Sons.

Sawyer, C. N., McCarty, A. L. & Parking, G. F. (1994). *Chemistry for Environmental Engineering.* New York: McGraw-Hill.

Smith, R-K. (1995). *Water and Wastewater Laboratory Techniques.* Alexandria, VA: Water Environment Federation.

Smith, R. L. (1974). *Ecology and Field Biology.* New York: Harper & Row, Pub.

Snoeyink, V. L. & Jenkins, D. (1988). *Water Chemistry.* 2nd ed. New York: John Wiley & Sons.

Spellman, F. R. (1996a). *Safe Work Practices for Wastewater Treatment Plants.* Lancaster, PA: Technomic Publishing Co., Inc.

Spellman, F. R. (1996b). *Stream Ecology and Self-Purification: An Introduction for Wastewater and Water Specialists.* Lancaster, PA: Technomic Publishing Co., Inc.

Spellman, F. R. (1997a). *Wastewater Biosolids to Compost.* Lancaster, PA: Technomic Publishing Co., Inc.

Spellman, F. R. (1997b). *Microbiology for Water/Wastewater Operators.* Lancaster, PA: Technomic Publishing Co., Inc.

Spellman, F. R. (1997c). *A Guide to Compliance for Process Safety Management/Risk Management Planning (PSM/RMP).* Lancaster, PA: Technomic Publishing Co., Inc.

Sterritt, R. M. & Lester, J. M. (1988). *Microbiology for Environmental and Public Health Engineers.* London: E. and F.N. Spoon.

Tchobanglous, G. & Schroeder, E. D. (1985). *Water Quality.* Reading, Massachusetts: Addison-Wesley Publishing Company.

The Virginian-Pilot (Norfolk), 24, January 1997, A6.

Turk, J. & Turk, A. (1988). *Environmental Science.* 4th ed. Philadelphia: Saunders College Publishing.

U.S. Bureau of Census (1987). *Statistical Abstract of the United States: 1988.* Washington, D.C.

U.S. Department of Health, Education and Welfare (1982). *Drinking Water Standards.* PHS Bulletin No. 956, Public Health Service.

USEPA (1975). National Interim Primary Drinking Water Regulations. *Federal Register,* Part IV, December.

USEPA (1992). *Guidelines for Water Reuse.* EPA/625/R-92/004.

U.S. Water Resources Council (1978). *The Nation's Water Resources, 1975–2000.* Second National Water Assessment, Washington, D.C., December.

van der Veen, C. & Graveland, A. (1988). Central Softening by Crystallization in a Fluidized-Bed Process, *Journal of American Water Works Association, 80,* 6, 51.

Watson, L. (1988). *The Water Planet: A Celebration of the Wonder of Water.* New York: Crown Publishers, Inc.

Weber, S. ed. (1988). *USA by Numbers.* Washington, DC: Zero Population Growth.

WEF (1992). *Nonpoint Source Pollution.* Alexandria, VA: Water Environment Federation.

Welch, E. B. (1980). *Ecological Effects of Waste Water.* Cambridge University Press, Cambridge.

WERF (1994). *Water Reuse.* Alexandria, VA: Water Environment Research Foundation.

Wetzel, R. G. (1983). *Limnology.* New York: Harcourt Brace Jovanovich College Publishers.

Wheeler, D. (1997). Regulated Community Viewpoint, Issues, and Concerns. Paper delivered at *51st Annual Virginia Water Environment Association Conference,* Roanoke, Virginia, May 4, 1997.

Wistreich, G. A. & Lechtman, M. D. (1980). *Microbiology,* 3rd ed. New York: Macmillan Publishing Co.

WRI & IIED (1986). *World Resources 1986.* New York: Basic Books.

WRI & IIED (1987). *World Resources 1987.* New York: Basic Books.

WRI & IIED (1988). *World Resources 1988–89.* New York: Basic Books.

Index